Welding Metallurgy of Aluminium Alloys

Welding Metallurgy of Aluminium Alloys

Design, Processes, and Simulations

Reza Beygi, Eduardo André de Sousa Marques, Ricardo João Camilo Carbas, and Lucas Filipe Martins da Silva

WILEY VCH

Authors

Prof. Reza Beygi
Arak University
Arak, Iran

Prof. Eduardo André de Sousa Marques
University of Porto
Rua Dr. Roberto Frias
4200-465 Porto
Portugal

Dr. Ricardo João Camilo Carbas
INEGI Institute of Science & Innovation
Rua Dr. Roberto Frias
4200-465 Porto
Portugal

Prof. Lucas Filipe Martins da Silva
University of Porto
Rua Dr. Roberto Frias
4200-465 Porto
Portugal

Cover Design: Wiley
Cover Image: © SimoneN/Shutterstock

Typesetting Straive, Chennai, India

Printed and bound by CPI Group (UK) Ltd, Croydon, CR0 4YY

C9783527353347_040825

Library of Congress Card No.: applied for

British Library Cataloguing-in-Publication Data A catalogue record for this book is available from the British Library.

Bibliographic information published by the Deutsche Nationalbibliothek
The Deutsche Nationalbibliothek lists this publication in the Deutsche Nationalbibliografie; detailed bibliographic data are available on the Internet at <http://dnb.d-nb.de>.

Print ISBN: 978-3-527-35334-7
ePDF ISBN: 978-3-527-84535-4
ePub ISBN: 978-3-527-84536-1
oBook ISBN: 978-3-527-84537-8

Contents

Preface

The field of welding metallurgy for aluminum alloys continues to evolve, addressing new challenges and meeting industry demands for advanced materials and methods. *Welding Metallurgy of Aluminium Alloys: Design, Processes, and Simulations* serves as a comprehensive guide for students, engineers, and professionals who seek a deep understanding of aluminum welding processes and the underlying principles. This book covers critical topics, from microstructure evolution during welding to defect mechanisms and prevention, offering both theoretical insights and practical applications.

Drawing on over 450 research articles, this book bridges the gap between fundamental science and industrial practice. Each chapter explores key aspects of welding metallurgy, providing detailed explanations of microstructural changes across various processes and an in-depth analysis of common defects – both technical and metallurgical. Particular attention is given to defects such as hot cracking, porosity, and softening, as well as process-specific issues such as lack of penetration and kissing bonds in friction stir welding. Mechanisms of defect formation and strategies for mitigation are discussed in depth, enabling improvements in weld quality and performance.

In addition to defect analysis, this book covers simulation approaches for each welding process, examining temperature distribution, thermal history, residual stress, and distortion. Special focus is placed on material flow in friction stir welding, while weld design principles are discussed for all processes, supporting researchers and practitioners in their work.

It is hoped that this book will serve as a valuable resource for those studying or working with aluminum welding, whether in academic settings or industrial applications. By presenting a detailed account of welding metallurgy for aluminum alloys, this book aims to contribute to a deeper understanding and advancement of this important field.

List of Abbreviations

AM	additive manufacturing
AC	alternative current
ALE	arbitrary Lagrangian–Eulerian
BTR	brittle temperature range
CGZ	coarse grain zone
CNT	carbon nanotube
CMT	cold metal transfer
CEL	coupled Eulerian–Lagrangian
DC	direct current
EBSD	electron back-scattered diffraction
EBW	electron beam welding
FE	finite element
FFT	Fast Fourier Transform
FSW	friction stir welding
FSP	friction stir processing
FZ	fusion zone
GMAW	gas metal arc welding
GTAW	gas tungsten arc welding
HAZ	heat-affected zone
HTA	heat-treatable aluminum
IMC	intermetallic compound
LAHW	laser arc hybrid welding
LBW	laser beam welding
LTW	laser transmission welding
LPBF	laser powder bed fusion
MFDC	medium frequency direct current
MIG	metal inert gas
NHTA	non-heat-treatable aluminum
NZ	nugget zone
PMZ	partially melted zone
PSN	particle stimulated nucleation
PLBW	pulsed laser beam welding
RSC	resistance spot clinching

RSW	resistance spot welding
RW	resistance welding
NZ	nugget zone
PAW	plasma arc welding
PFZ	precipitate free zone
PWHT	post-weld heat treatment
SEM	scanning electron microscopy
SC	solidification cracking
SCS	solidification cracking susceptibility
SMAW	shielded metal arc welding
TEM	transmission electron microscopy
TMAZ	thermo-mechanically affected zone
TIG	tungsten inert gas
UTS	ultimate tensile strength
VP	variable polarity
YS	yield strength

1

Aluminum Alloys: Fundamentals of Welding Metallurgy

1.1 Introduction

Aluminum (Al) alloys have been used traditionally for structural parts of aircrafts. At the beginning of the twenty-first century, the leading aircraft companies started to use composite materials due to their advantages such as low weight. Nonetheless, the usage of Al in structural parts still continues due to better recyclability of these materials and satisfying the weight reduction of the structure [1]. Al is a light alloy with a density of 2.7 g/cm^3, is nonmagnetic, has an FCC crystal structure, has high formability and low-temperature toughness, and has a relatively high corrosion resistance depending on the alloying elements [2]. Al alloys are also easily processible with casting, machining, and extrusion. Some Al alloys capable of precipitation hardening have strengths comparable to steels [2]. In contrast to carbon and low-alloy steels, Al does not possess any polymorph; therefore, only the solidification behavior determines the properties of the weld zone in the fusion welding [3]. Cooling of the solidified weld zone does not have a prominent effect on the microstructure of the weld due to absence of any polymorph transformation in Al alloys.

Al alloys are classified based on their mechanical properties, manufacturing process, and alloying elements. Wrought Al alloys are determined by the degree of cold work and chemical composition. Cast Al alloys are determined by the casting process and chemical composition. Other mechanisms such as precipitation hardening and solid-solution hardening may also contribute to the strength of Al alloys [2]. The alloying elements that promote precipitation hardening in Al are Cu, Cu—Mg, Cu—Li, Cu—Si, Zn, Zn—Mg, Zn—Mg—Cu, and Li—Cu—Mg [2]. The alloying elements that promote solid-solution hardening are Mn, Si, Zn, Mg, and Cu [2].

The designation system for wrought Al alloys in the European standard system is a four-digit numeric. Figure 1.1 shows the designation system according to the European standard. There is also a supplementary designation system whose basis is a chemical symbol. The applications of some Al alloys are provided in Table 1.1.

Along with austenitic stainless steel and Ni alloys, Al alloys are good candidates for the construction of LNG cargo containment, as Al possesses a high toughness due to its FCC structure [11]. The usage of Al alloys as structural materials requires their joining in various joint designs. Mechanical joining methods such as riveting have been used in lap configuration, which adds to the weight of the

Welding Metallurgy of Aluminium Alloys: Design, Processes, and Simulations, First Edition.
Reza Beygi, Eduardo André de Sousa Marques, Ricardo João Camilo Carbas, and Lucas Filipe Martins da Silva.

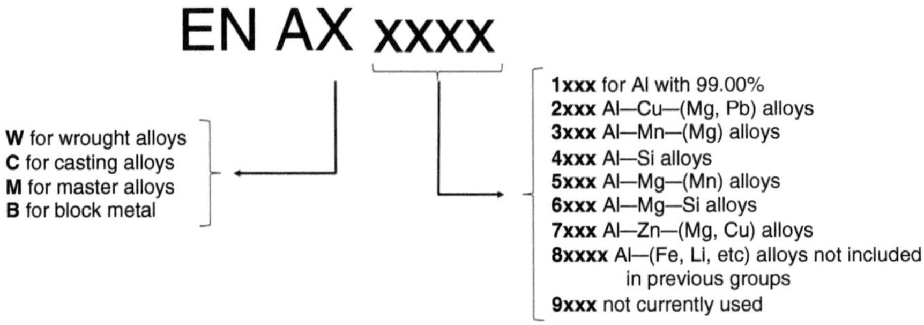

Figure 1.1 Designation system for wrought aluminum alloys according to the European standard.

Table 1.1 The applications of some Al alloys.

Alloy	Application
AA2xxx	Propellant cabin of launch vehicles [4]
AA5xxx	Shipbuilding [5]
AA6xxx	Car body [6], battery tray [7]
AA7xxx	Bicycle frame [8], panel structure of launcher [9], underframe in high-speed trains [10]

structure. An example is the stringer–skin connection used in the fuselage of the aircraft (Figure 1.2a), where additional adhesive is used to seal the joint. Welding technologies help to reduce the weight by eliminating the overlap region and the sealant and while maintaining the same performance (Figure 1.2b) [12]. In Tee-joining, the high heat conductivity of Al causes a high thermal gradient to form between the weld pool and base material. Lap joining also adds to the weight of the structure and hence is used for thin sheets less than 1 mm, where welding processes such as resistance welding are to be used. Regarding the weld design, fillet welds are usually avoided in the structural parts under fatigue due to crack initiation from the root of the weld [13]. The joint design also affects the residual stress during welding. For instance, the Tee-joints cause a higher residual stress during welding than lap joints due to higher heat sink effect in the Tee-joints, which causes a higher cooling rate [14].

The Al alloys that are heat-treatable (HTA) need to undergo specific sequences during forming and assembling to obtain desired mechanical properties. For example, in the car body, the Al sheets are welded in a solution-treated or hot-formed state. The final aging treatment, which is connected to the paint bake cycle, is performed after welding [15]. Figure 1.3 shows the schematic workflow of Al alloy used in the car body. Hot forming of Al alloy is performed in the solid-solution state, which is quenched subsequently to preserve the solid-solution state. The joining methods are performed after this stage, and then a paint bake cycle is carried out to regain the strength through aging treatment.

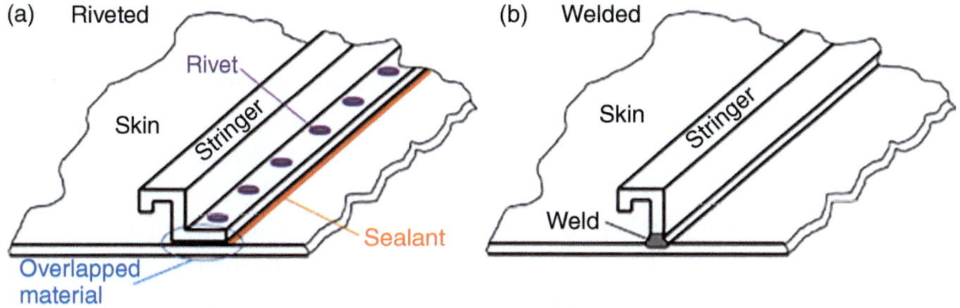

Figure 1.2 Joining of a stringer–skin aluminum part used in the fuselage of the aircraft made by (a) riveting and (b) welding. Source: Ref. [12]/with permission of Elsevier.

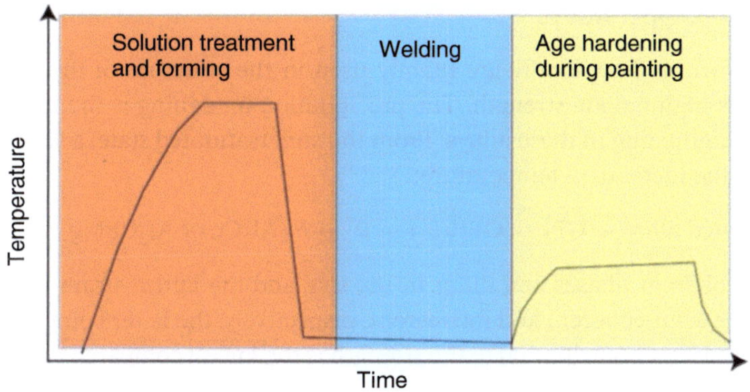

Figure 1.3 The forming, solution treatment, welding, and age hardening while paint baking of Al alloys used in the car body.

There are some challenges during welding of Al alloys, and understanding their mechanisms helps to avoid them and hence obtain the optimum performance of the welded structure. Some issues that exist during welding steel are not present during welding Al due to its physical properties. For instance, Al is a nonmagnetic material and therefore the problems of arc blowing will not occur during arc welding processes. Instead, other issues are pronounced during welding Al. The thermal conductivity of Al is six times that of steel and therefore the welding processes with high power density need to be used for Al. Though the melting point of Al is low, due to high specific heat of Al the heat sources need to have high intensity. Due to the high heat conductivity of Al, the use of welding processes with low heat intensity for wrought alloys and precipitation-hardened alloys produces a wide heat-affected zone (HAZ) with lower strength (due to softening) and distortion. In precipitation-hardened alloys, it is also probable that the precipitates in the HAZ dissolve and reprecipitate, which leads to brittleness [2]. The thermal expansion coefficient of Al is twice that of steel. This along with its low Young's modulus causes a high distortion, especially in thin structures, making it difficult to maintain the tolerances [16]. To keep the distortion within the limits of tolerance, the components need to be tightly clamped and often tack-welded. The welding sequence also needs to be planned carefully [2].

A significant difference between steel and Al arises from metallurgical aspects. Structural steels encounter phase transformation during cooling; the most important one is the austenite to ferrite transformation. While the strength of structural steels increases in the HAZ during welding, Al alloys lose their strength in the HAZ. A higher number of issues exist while welding Al alloys that are heat-treatable, which means their strength is increased by precipitation. Other Al alloys get their strength by other mechanisms such as solution hardening and work hardening. Three main series of heat-treatable Al alloys are introduced in the following.

1.1.1 Aluminum–Copper Alloys

Al—Cu alloys known as 2xxx family are mainly used in the structure of the air-frames and possess high fatigue strength. The precipitation hardening is the main mechanism of strengthening in these alloys. From the supersaturated state, a series of phases form by the increase in temperature:

$$\text{Supersaturated alloy} \rightarrow \text{GPI} \rightarrow \text{GPII}(\theta') \rightarrow \theta' \rightarrow \theta\ (\text{Al}_2\text{Cu or Al}_2\text{CuMg})$$

GP phases are coherent phases and differ in the size and the lattice strain they induce. θ' and θ are semi-coherent and incoherent, respectively, the latter being the equilibrium phase, whose contribution to the strength is the lowest compared to the other phases.

1.1.2 Aluminum–Zinc–Magnesium Alloys

$MgZn_2$ is the main precipitate in this group of Al alloy. These Al alloys are less sensitive to quenching, and thus a supersaturated condition can be obtained even after air cooling. Natural aging can occur in these alloys after a couple of months to recover their initial hardness, making these alloys to have a self-hardening effect [17]. So, these alloys do not need a post-weld heat treatment (PWHT) to recover their hardness. However, these alloys are susceptible to solidification cracking (SC) in the fusion zone (FZ) and liquidation cracking in HAZ. Evaporation of Mg and Zn due to their low boiling point can cause pore formation during welding [18].

1.1.3 Aluminum–Magnesium–Silicium

The strength in the HAZ of these alloys is only partially restored by subsequent aging, as explained in Figure 1.28. The degree of strength lost in the HAZ depends on the heat and the time it receives this heat. It means that the welding process and its parameters determine the degree of strength lost in the HAZ. A lower heat input favors a narrower HAZ region that is less softened.

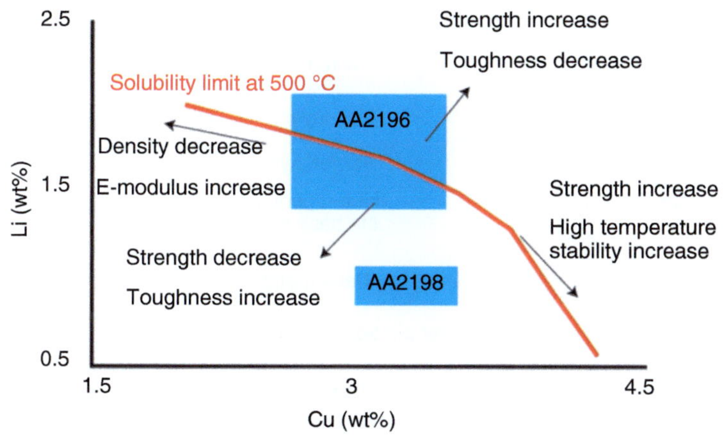

Figure 1.4 Schematic representation of mechanical properties of Al—Li—Cu alloys in relation to Cu and Li content. Source: Adapted from [19].

1.1.4 Aluminum–Lithium Alloys

These alloys, as special Al alloys, are used in the aircraft industry. Lithium has a low density and a high solubility in Al and hence offers a great potential to be used in the structure of airplanes. 1% lithium decreases the density of Al by 3% and increases the elastic modulus by 6% [19]. Some Al-Li alloys belong to Al—Cu—Li alloys such as AA2196 (found as extruded parts) and AA2198 (found as sheets and plates) [19]. The mechanical properties of these alloys primarily depend on the content of copper and lithium, as shown in Figure 1.4 [20]. The red line shows the solubility limit at 500 °C. The increase in Cu leads to an increase in Al_2CuLi precipitation hardening phases, which increase the strength. An increase in Li promotes the formation of Al_3Li precipitates and lowers the density. In addition to precipitates, the dispersoids may also exist. Dispersoids, in contrast to precipitates, are formed above the solubility limit temperature, and hence they cannot be altered by heat treatment of the solid alloy. Manganese is an element that forms incoherent dispersoids and increases the toughness and fatigue strength [19]. In Al—Mg—Li alloys, the main precipitate is Al_3Li. Al—Li alloys are prone to pore formation during welding, which downgrades their weldability [21].

1.2 Weldability

Welding of Al affects the mechanical properties due to the change in the microstructure during solidification as well as the heat effects in the solid state. The surface tension and the viscosity of the Al melt are low [22]. This increases the risk of melt burn through and spatter during welding Al. The physical, chemical, and metallurgical properties of Al alloys influence their behavior during the welding processes. Despite having a low melting temperature (up to 650 °C), a high thermal conductivity of Al ($\lambda = 120 - 240$ W/m K) necessitates a high heat intensity for welding.

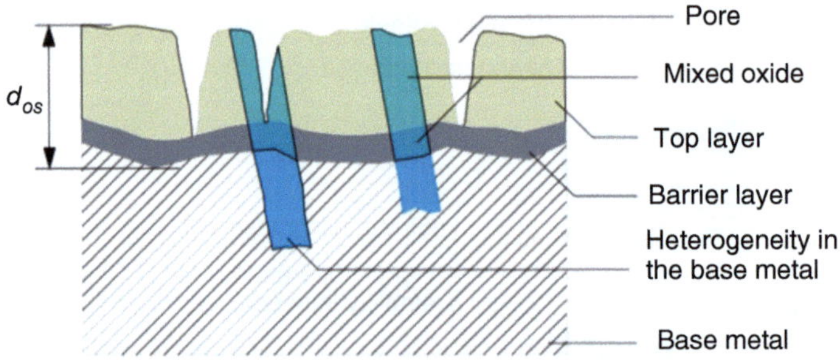

Figure 1.5 Structure of the Al oxide layer in wet air. Source: Adapted from [19].

The thermal conductivity of Al decreases with increasing temperature. Some Al alloys, despite having excellent properties, exhibit low joint efficiency by welding and hence their usage is less favored than other Al alloys that have lower mechanical properties but exhibit better joint efficiency. Some features of Al make it different from steels in terms of weldability. The first feature is the oxide of Al alloy. The special characteristics of Al oxide necessitate some considerations to obtain weldability. The presence of hydrogen is another factor that causes weldability challenges. This is very different from steel in which hydrogen causes completely different problems in welding. Hot cracking is another problem that occurs in the FZ during the final stage of solidification. A similar phenomenon named liquation cracking occurs adjacent to the FZ in partially melted zone (PMZ), which degrades the mechanical properties. The mechanisms of all these phenomena are explained in detail for Al alloys, whose understanding helps to avoid them during welding and improve the weldability.

1.2.1 Aluminum Oxide

The oxide of aluminum (Al_2O_3) is very dense, tough, and thermally stable and has a high melting temperature ($T_m = 2050\,°C$). It is also nonconductive to the electricity. The oxide layer consists of a narrow barrier layer and a thick layer, which have a certain porosity. The presence of alloying elements promotes the formation of complex oxides as well as the change in the thickness of the oxide layer. This oxide layer can trap hydrogen, which changes the characteristics of the oxide layer [19]. The scheme of the oxide layer on the surface of Al in the wet air is shown in Figure 1.5.

The oxide layer can cause an incomplete bonding by inhibiting a metallurgical bonding during welding if it is not removed away by proper methods. In arc welding processes this is done by cleaning effect of the arc in reverse polarity, during which the workpiece is in the positive pole. The oxide layer also influences the plasma spectra and its electronic temperature during arc or laser welding processes [23]. In the absence of the oxide film, the elements are more easily heated to the ionization temperature. Also, the electronic temperature of the plasma is higher, which is due to the absorption of the heat by the oxide film. This causes the depth and the width of the weld to be lower in the presence of the oxide film [23].

Due to its higher melting temperature, the oxide is not melted during welding Al and covers the Al surface with a firm shell, which interferes with the formation of the weld pool [24]. Due to its detrimental effect, the oxide layer has to be removed before welding.

1.2.2 Hydrogen and Pore Formation

Hydrogen is the main gas that forms pores during welding Al alloys. The elements of nitrogen and oxygen do not cause pore formation, as they have a high affinity to make stable nitrides and oxides and therefore are not available in the form of gas. Hydrogen is soluble to a high extent in molten Al, whose solubility decreases abruptly after solidification (Figure 1.6). After solidification, the excess amount of hydrogen, which is not soluble in the solid state, will exit and stay in the melt as pores. This problem is exacerbated when the cooling rate is rapid. By increasing the welding speed, the gas escape is hindered and the pores will become higher in quantity. The position of the welding is also determinant in pore formation. The gas can leave the melt pool easily when welding is performed in the flat position. In the overhead position, there is no escape path for the gas bubbles, and therefore, this position of welding should be avoided for welding Al. One source of hydrogen is humidity, by which the water reacts with Al in the arc space and consequently hydrogen is released according to:

$$2Al + 3H_2O \rightarrow Al_2O_3 + 3H_2 \quad (1.1)$$

The sources of hydrogen are the humidity in the protecting gas, the humidity entered from the atmosphere, and the residues on the surface of the workpiece or the filler, such as paint, oil, and hydroxides.

Another source of hydrogen is the oxide layer of the Al. The pores that result from the oxide layer of the base material appear as uniform coarse pores over the weld length. The pores that result from the hydrogen of the oxide layer of the filler material appear as uniform fine pores [19]. The thicker the oxide layer, the higher the amount of hydrogen. The thickness of the oxide layer, as mentioned before, depends on the alloying elements or, in other words, the type of Al alloy. For example, AA2198 possesses a thinner oxide layer than AA2196. Another factor that determines the amount of hydrogen from the oxide layer is the thickness of the Al

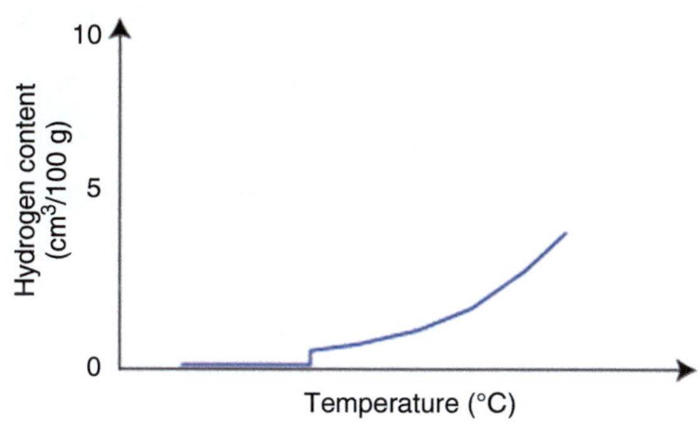

Figure 1.6 Solubility of hydrogen in aluminum at various temperatures.

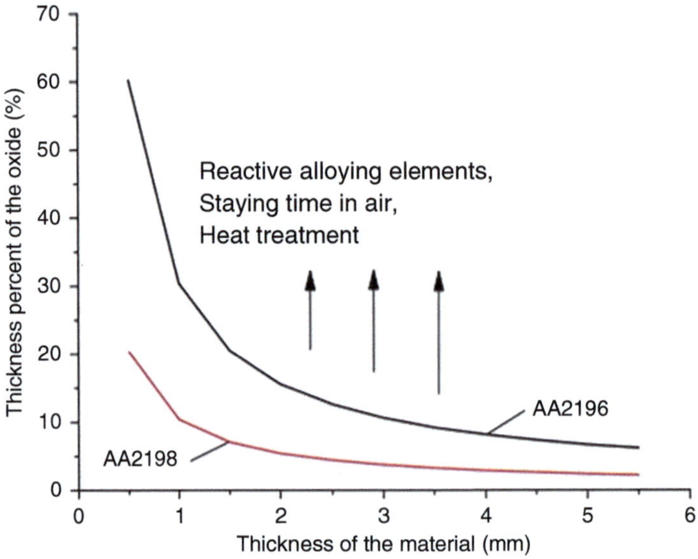

Figure 1.7 Relation between the portion of the oxide layer and the thickness of the sheet for two Al alloys. Source: Adapted from [19].

sheet [19]. By increasing the thickness of the Al sheet, the ratio of the oxide/total mass of Al decreases, which in return lowers the amount of hydrogen introduced. The portion of the oxide layer decreases exponentially with thickness, as shown schematically in Figure 1.7. The annealing temperature and time can also increase the thickness of the oxide layer.

The oxide layer can also produce pores in another way. During laser welding of Al, the high temperature inside the keyhole vaporizes the oxide layer at the interface, which appears as pores (red squares in Figure 1.8) [25]. These pores are distinguished by the pores caused by the keyhole instability, which form at the bottom of the weld (blue squares in Figure 1.8). The mechanism of pore formation by keyhole is explained in detail in Chapter 3. Also, the mechanism that controls these kinds of pores will be discussed. As an example, introducing a gap between

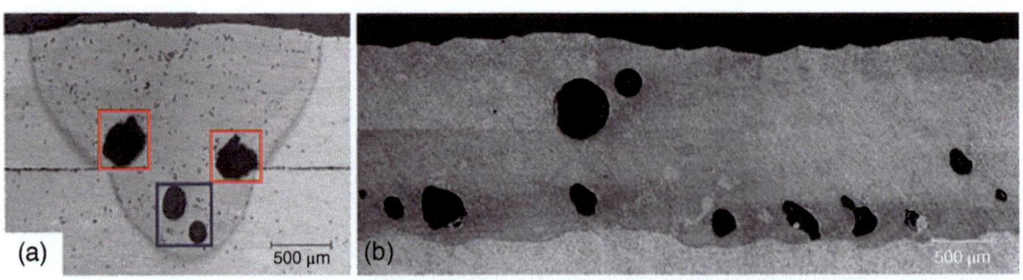

Figure 1.8 Pores caused by oxide layer (inside the red squares) during laser welding of a 6061 alloy. (a) Cross section view and (b) longitudinal view. Source: Ref. [25]/with permission of Elsevier.

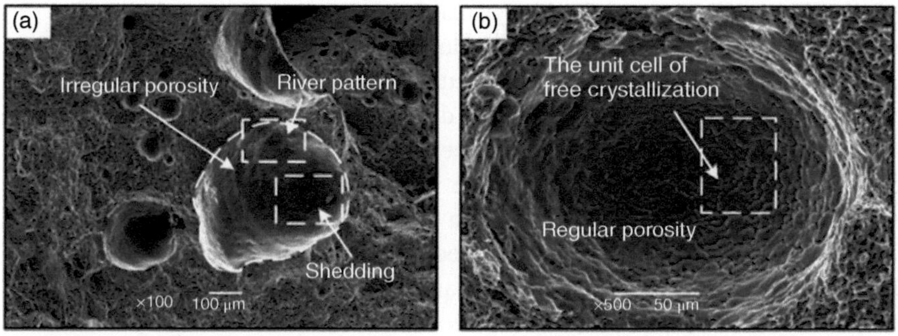

Figure 1.9 SEM image of (a) technical or keyhole-induced pore and (b) metallurgical pore. Source: Ref. [27]/Taylor & Francis.

the plates provides a vent to relieve the pressure, causing a more stable keyhole [25]. This leads to less pore formation and a stronger weld.

Based on these, the surface of Al needs to be cleaned before welding either by mechanical or chemical methods [19]. The degree of removal of the surface oxide depends on the thickness of the oxide layer. Considering that the thickness of the oxide is not uniform (see Figure 1.5), the removal of the surface should make sure that the oxide layer is removed completely. The mechanical methods of removing the oxide layer, especially the milling processes, are the most proper ones. Though in these methods the surface becomes wavy, it is free of adhered deposits, which appear after chemical methods [2]. One example of the chemical cleaning method is soaking in 20% sodium hydroxide solution and subsequent soaking in dilute nitric acid to neutralize the residual alkali before welding [26]. Due to the rapid formation of the oxide layer, even in the vacuum, the welding process should be carried out right after the surface treatment.

In fusion welding processes wherein a keyhole exists, such as laser beam welding, another kind of pores can form during welding. The source of these pores is not hydrogen, but the metallic vapor and the shielding gas entrapped by the keyhole collapse. When the melt pool solidifies rapidly, these entrapped vapors and gases do not have enough time to escape from the melt pool, forming a keyhole-induced pore [26]. A scanning electron microscopy (SEM) image of this hole is provided in Figure 1.9a. These pores have irregular shapes with step-like inner walls and traces of solidification contraction [26]. The pores that are caused by hydrogen or by metal vapor produced by the burning of the metal elements are referred to as metallurgical pores [27]. The metallurgical pores have regular shapes, and the inner interface is smooth due to free crystallization cell boundary on the wall of the pore. The internal pressure of accumulated hydrogen inside the pore in the solidification stage initiates after nucleation of the pore and then the continuous diffusion of hydrogen into it [27]. The pores formed by keyhole collapse in the unstable welding pool are also referred to as the technical pores, which are irregular and large [27]. These pores can be reduced by controlling the process parameters. The pores caused by hydrogen are smaller in size than the ones induced by instability of the keyhole [28]. The details of the formation of the technical pores are addressed in Chapter 3.

1.2.2.1 Effects of Porosity on the Mechanical Behavior

The porosity in the weld metal of Al reduces the ductility, tensile strength, and fatigue stress [29]. Deformation concentrates around the porosities, and a higher strain rate exists in the regions with these defects, which subsequently reduces the ductility [30]. If the porosity percent is lower than 4% volume, it does not influence the static mechanical behavior [31]. The pores exhibit on the fracture surface, showing their contribution to the initiation of the fracture [32].

The fatigue strength will not be influenced if the diameter of the pores is less than 0.2 mm, as these pores are present in globular form. By increasing the diameter of the pores, the dynamic behavior will be influenced. The dispersion of the pores is also important in the determination of the mechanical behavior. A solitary scattered pore or pore nest is less detrimental than line pores with very small sizes. In the welded specimens without any pores, the fatigue crack initiates at the surface. While in the presence of porosity, the fatigue cracks initiate near the edge of the porosities, and by increasing the porosity, the fatigue stress reduces [29, 33].

The pores are not the only cause of degradation of the mechanical properties. The cracks made by metallurgical phenomena such as hot cracking that occurs during solidification also have an impact. The existence of various defects such as pores and cracks in the FZ is reported in the literature [34]. Understanding the mechanism of formation of these cracks is necessary to avoid them during welding.

1.2.3 Hot Cracking

SC is a very typical failure in Al alloys. In contrast to pure metal, alloys solidify within a temperature range where both solid and liquid phases are present. A simple explanation of this crack is provided in Figure 1.10. The occurrence of SC is accompanied by local stress or strain as well as eutectic compounds with low melting temperatures present in the grain boundaries between the growing dendrites. Usually, Al alloys are more susceptible to this cracking than pure Al [35]. At the final stage of the solidification process, while the grains are solidified the eutectic phase is present in the form of liquid between the grains. The shrinkage of solidification causes the just solidified grains to move with respect to each other whose boundaries are filled with this liquid. This causes a separation of the grains, which, by further cooling and shrinkage, leads to the formation of cracks lined up together (Figure 1.10a). When enough liquid as well as a path for this liquid to fill these cracks are available, a healing process occurs that eliminates SC, as shown in Figure 1.10b.

Various theories have been proposed to address the details of hot cracking. During solidification when cooling down the melt, there is a slurry stage where both solid and liquid phases are present and the dendrite arms of the solid phase are not coalescent. The temperature is between the temperature of coherency and the liquidus ($T_{coh} < T < T_l$). By further cooling, when the temperature is below the coherency temperature and above the solidus temperature, the dendrites become coalescent and form a coherent solid network named mushy zone ($T_s < T < T_{coh}$) [36]. The schematics of these regions (slurry and mushy zones) are shown in Figure 1.11.

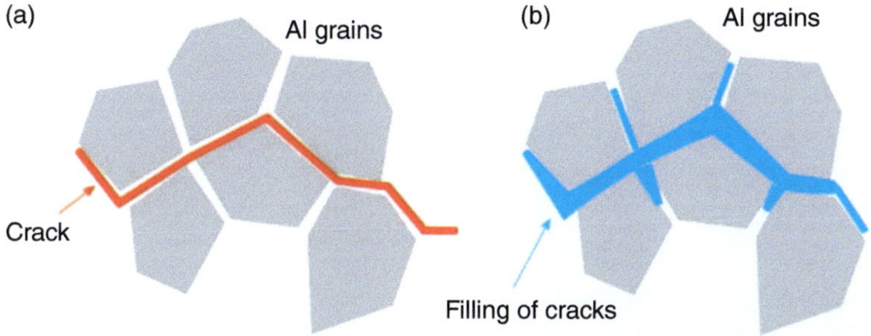

Figure 1.10 (a) The contribution of eutectic compounds at the grain boundaries to the formation of solidification cracking. (b) Filling of the crack by available eutectic compound.

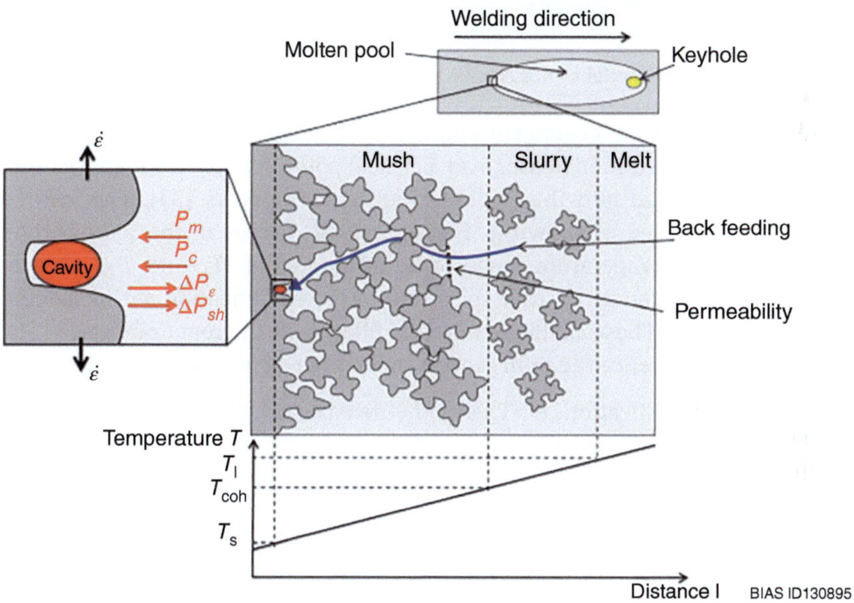

Figure 1.11 Schematic diagram of the semisolid molten pool. Source: Ref. [37]/Springer Nature.

According to "shrinkage-brittleness theory," when the dendrites grow during cooling, they form a network of solid structures below the coherence temperature, which is above the solidus temperature. By further decreasing the temperature below the coherence temperature, a high strain rate develops in the solid structure, which may cause the dendrite arms to open and get separated in the coherent zone. The remaining liquid may not penetrate the bridges between the solids due to low permeability of the solid and high viscosity of the liquid. This process continues until the solidus temperature is reached. The interval between the coherence temperature and the solidus temperature is named the effective interval, which is referred to as the brittle temperature range (BTR). At the starting point of BTR, the dendrites start to connect and hence the thermal stresses form. At the end

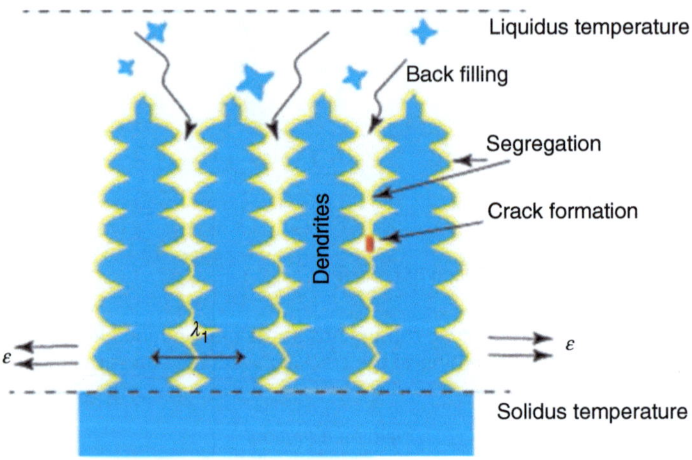

Figure 1.12 Schematic model of hot cracking and its influencing variables.

of BTR, the dendrites are coalescent and hence this region gains a strength [38]. When the shrinkage strain increases over a critical point in BTR, this region cannot bear the resultant load and thus solidification crack appears [39]. This can be schematically explained as shown in Figure 1.12. SC occurs in the solidification interval (ψ) in the temperature range of liquid and solid (T_L and T_S) wherein cavities in the inter-dendritic space form due to the solidification shrinkage and thermal contraction. These cavities cannot be filled with the remaining melt. The following factors influence the refill process of the inter-dendritic space [40]:

- The length of solidification, ψ: This is the distance which the remaining melt should travel to fill the cavity.
- Solidification time of the weld pool: It is the available time span for the remaining melt to flow to the cavities.
- Permeability of the melt in the solidification interval: It shows how easily the remaining melt flows between the dendrites. After a certain distance, the inter-dendritic space becomes too small to allow for the remaining melt to flow in. This can be metallographically quantified as the primary dendritic or cell-arms distance (λ_1). The permeability of a dendritic network (k_{dend}) is obtained from [41]:

$$k_{dend} = \frac{f_l^3 \lambda_2^2}{180 f_s^2} \tag{1.2}$$

  where λ_2, f_s, and f_l are the secondary dendrite arm spacing, the fraction of solid, and the fraction of liquid, respectively.
- Segregation in the inter-dendritic network: Through this, the remaining melt is enriched with low-melting eutectics with a wide solidification range.
- The strain and strain rate, which act on the thin liquid film between the dendrites and cause the dendrites to open cracks.
- The tensile stresses, which develop in the coalescent part of the two-phase region (solid–liquid) where the refill of the remaining melt cannot occur.

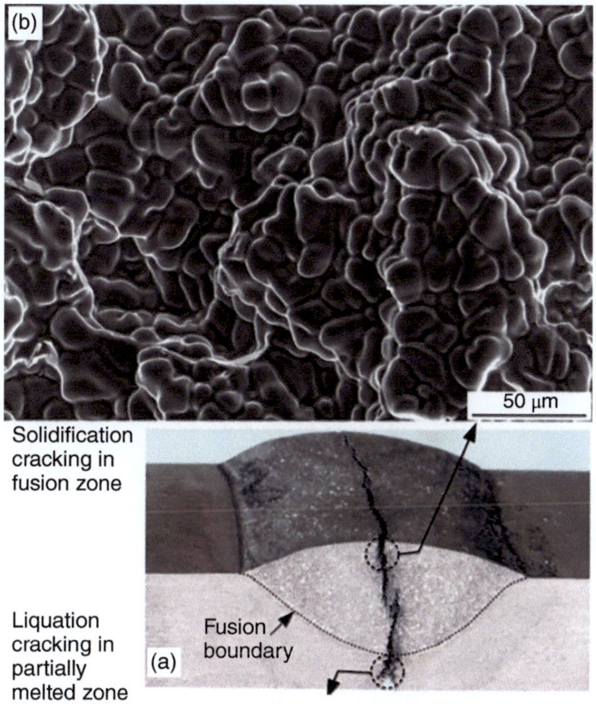

Figure 1.13 Fracture surface of an SC developed during welding of an AA6061. (a) Macro image of the weld. (b) SEM image of the crack surface. Source: Ref. [43]/Springer Nature.

SC can be distinguished in several ways. Usually, it occurs at the center of the weld pool before the weld is cooled down. When a weld with an SC is loaded to failure, the fracture surface analysis by SEM can reveal some unique characteristics of the SC. The main feature of SC is a free solidified fracture surface on which low melting eutectic is present (Figure 1.13) [42]. Along with SC, another form of hot cracking, called liquation cracking, may occur, which is explained at the end of this section.

The commercial pure Al is not susceptible to SC due to the absence of the solidification temperature range [35]. The composition of the melt zone is the main parameter that determines the occurrence of SC. The binary systems of Al alloys such as Al—Si, Al—Cu, and Al—Mg are susceptible to hot cracking in specific ranges of composition. SC is more likely to occur in multi-element Al alloys, for example precipitation-hardened alloys, as the number of compounds that promote hot cracking increases. In fact, the alloying elements promote the segregation during dendrite growth, which appear as low-melting compounds between the dendrites. SC, as a metallurgical defect, occurs during cooling in the range of liquidus–solidus temperature range under residual stress in the weld zone and in the weld crater in the presence of the alloying elements. The larger the solidification interval and the higher the residual stress, the more probable the occurrence of SC. The effect of the solidification interval on the hot crack tendency is schematically shown in Figure 1.14. This kind of diagram gives some important criteria that can be used to assess the SC sensitivity. The dwelling time in the interval and the span of the concentration ($\Delta C = C_{max} - C_1$) are two features that yield good measures of the segregation and

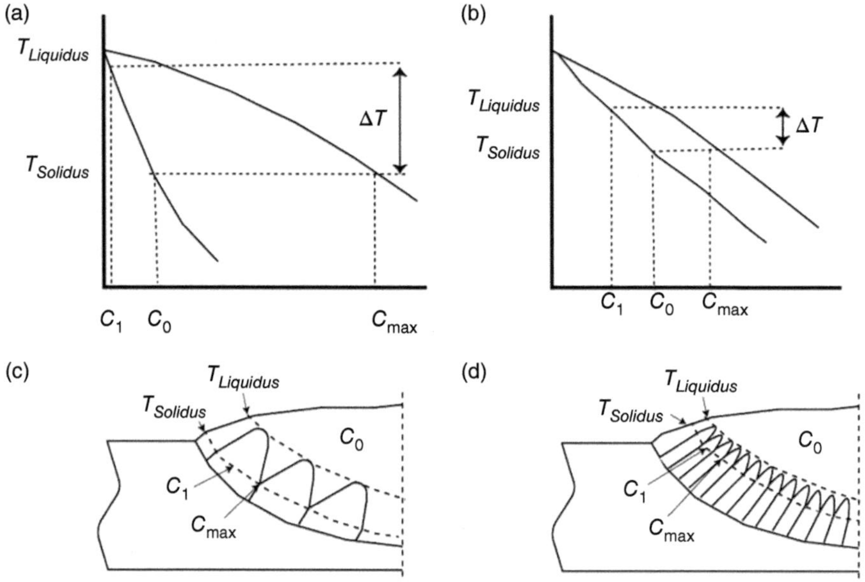

Figure 1.14 Effect of (a and b) solidification interval (ΔT) on (c and d) the segregation of elements at the center of the weld.

consequent SC. When ΔT is large (Figure 1.14a, c), the segregation is pronounced, and when ΔT is small (Figure 1.14b, d), the segregation is low. The liquidus and solidus temperatures can be calculated by the silicon equivalent concept, whereby the chemical composition of a multi-element weld was simplified using an Al/Si binary system. In this way, the silicon equivalent of each element ($Si_{eq}^{X_i}$) is obtained by its corresponding equation, and subsequently, the total Si equivalent of the melt (Si_{eq}) can be obtained by the following relationship [37]:

$$Si_{eq} = [Si] + \sum Si_{eq}^{X_i} \tag{1.3}$$

The liquidus and solidus temperatures and hence ΔT can be easily obtained by knowing this Si_{eq}.

When Al is alloyed above the maximum solubility, eutectic microstructure forms, which reduces the hot crack susceptibility. The results of the evaluation of the hot crack susceptibility of an AA6xxx alloy (Al—Mg—Si system) by ring casting samples [44] are shown in Figure 1.15. The increase in the alloying elements reduces the crack length and thus the SC susceptibility. The chemical composition and the resulting solidification range cannot always be solely used as a criterion for prediction of SC. The situation was reported in which some solidification cracks were present locally, despite being in the safe composition range. This indicates that other factors are needed to be taken into account. The induced strain rate during solidification plays an important role in SC occurrence [45]. The strain rate is dependent on the thermal gradient and cooling rate as well as the speed of backfilling the cracks. A two-level hot crack index was proposed by which both the composition and the thermal effects of the process are considered [46]. The shrinkage during solidification and thermal contraction on the one hand develop stress that promotes SC. On the other hand, they induce a pressure drop between the solidified dendrites, which causes the liquid to flow against the solidification direction. This makes a backfilling whose cause is a pressure difference [36]. This pressure drop (Δp_c), also called cavitation depression, which is caused by shrinkage (Δp_{sh}) and deformation (Δp_{def}), is obtained through [46]:

$$\Delta p_c = \Delta p_{sh} + \Delta p_{def} = \frac{180}{\lambda_2^2}\frac{v_T \beta \mu}{G} A + \frac{180}{\lambda_2^2}\frac{(1+\beta)\mu \dot{\varepsilon}(T)}{G^2} B \tag{1.4}$$

where λ_2, v_T, β, μ, G, and $\dot{\varepsilon}(T)$ are the secondary dendrite arm spacing, the velocity of isotherms, the shrinkage factor, the liquid viscosity, the thermal gradient, and the strain rate of deformation, respectively. A and B are obtained from [46]:

$$A = \int_{T_s}^{T_l} \frac{f_s(T)^2}{(1 - f_s(T))^2}\, dT$$

$$B = \int_{T_s}^{T_l} \frac{f_s(T)^2 \int_{T_s}^{T_1} f_s(T) dT}{(1 - f_s(T))^3}\, dT \tag{1.5}$$

where $f_s(T)$, T_l, and T_s are the fraction of solid at temperature T, the liquidus temperature, and the coherency temperature, respectively. The larger the parameters A and B, the larger the pressure drop and the higher the tendency for hot cracking [46]. Parameter A depends on the solidification temperature range and can be used as a criterion to predict the hot cracking susceptibility based on the chemical composition. The Rappaz Drezet Gremaud (RDG) theory is based on the critical depression criterion, above which the strain rate is high enough to induce cracking in the coherent zone [47]. The hot cracking susceptibility depends on the maximum strain rate that can be endured in the mushy zone and is obtained from:

$$\dot{\varepsilon}_{p,\max} = \frac{\lambda_2^2}{180}\frac{G}{(1+\beta)\mu}\Delta p_c - v_T \frac{\beta}{1+\beta} A \tag{1.6}$$

The higher $\dot{\varepsilon}_{p,\max}$, the lower the hot cracking sensitivity, as the higher strain rate reduces the critical strain at which the crack initiates [48]. This equation considers the effect of both the chemical composition (through parameter A) and the thermal effects of the process (through parameters (G, λ_2) [46]. Positive strain rate refers to a tensile load, and negative strain rate refers to a compressive load. When the critical strain rate is negative, it means that the strain developed during solidification should be compressive to avoid SC. Usually, as the stresses developed during solidification are tensile, the strain rate would also be tensile, meaning that the strain rate cannot be smaller than the critical strain rate and therefore the risk of SC is very high [49]. By choosing a proper filler material and dilution, the critical strain rate can be improved to avoid SC. An example is welding of AA606 with AA4043 filler material by which the critical strain rate was increased from −0.16%/s at 0% dilution to 0.35%/s at 16% dilution of AA4043 filler material [50]. The role of Si in this case is to refine the grains and change the bead shape, which both improve the SC.

There are factors other than the chemical composition and the welding parameters that affect the SC behavior. Figure 1.16 shows how the weld seam shape in

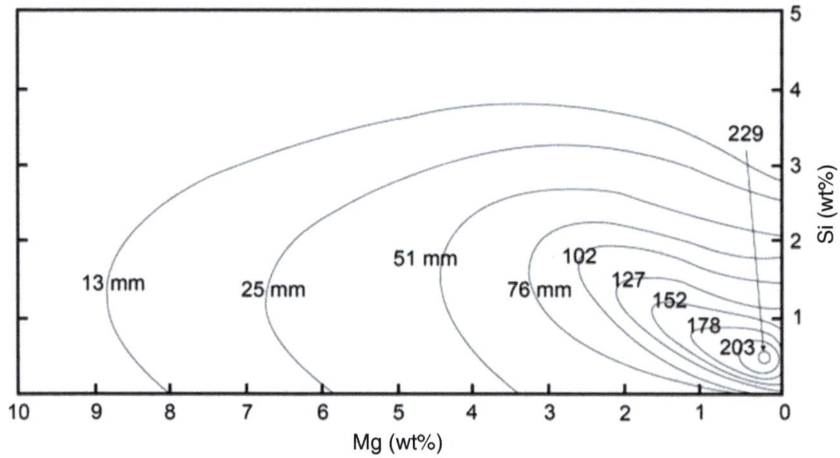

Figure 1.15 Length of the solidification crack with respect to the amount of Si and Mg for AA6xxx. Source: Ref. [40]/with permission of Dr. Martin Bielenin.

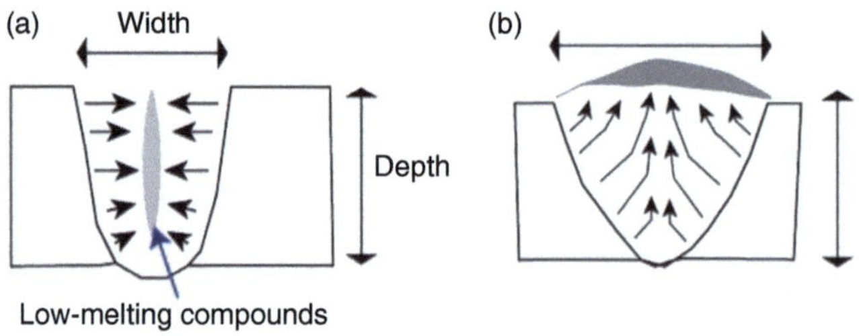

Figure 1.16 Effect of weld pool shape on the segregation and solidification cracking in a single-pass weld with (a) a high depth-to-width ratio and (b) a low depth-to-width ratio.

a single-pass weld affects the solidification and promotes SC. In fact, the shape and the size of the weld zone cause the low-melting compounds to segregate in the solidification front whose edges are parallel on both sides. Deep weld pools are prone to this type of SC. For example, in laser welding by which a narrow weld zone is created and the distance from the weld center line to the fusion line is small, there is a high susceptibility to SC due to this pronounced segregation [51]. The effect of depth-to-width ratio on the segregation and SC is shown in Figure 1.16.

Welding speed and heat input affect the shape of the weld pool and hence the SC. When the welding speed is low, the weld pool solidifies in a curvature toward the weld line, pushing the low-melt compounds to the weld pool. When the welding speed is high, the solidification fronts grow parallel and the low-melt compounds segregate at the center of the weld promoting SC (Figure 1.17).

Therefore, the following parameters can be controlled to lower the tendency of SC [2]:

- The increase in the number of passes and consequent decrease in the size of each pass decrease the SC susceptibility.

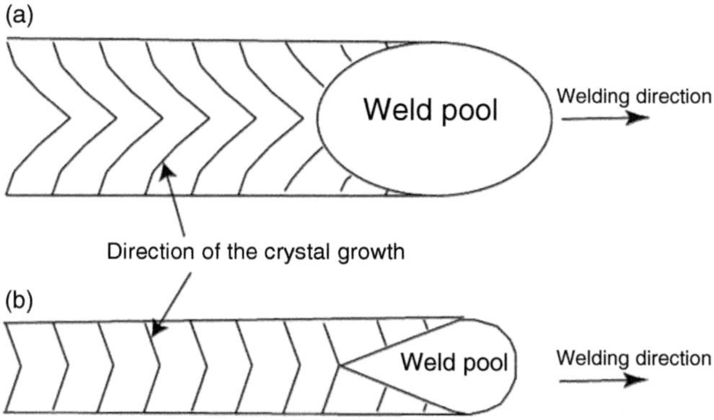

Figure 1.17 Effect of the welding speed on the solidification cracking at (a) low welding speed and (b) high welding speed.

- The low-melting compounds are formed from impurities, which usually are introduced from the base material. So, lowering the dilution decreases the percentage of impurities in the FZ and lowers the tendency for SC. The low dilution can be obtained by proper selection of welding process, proper selection of the welding parameters, and the weld seam preparation. For example, V-grooves result in lower dilution than square edges.
- Lower welding speed causes the solidification fronts to be curved at the center, leading to pushing off the impurities into the weld pool and weld surface as slag. Hence, the tendency for SC decreases.

Another effective method to alleviate the solidification crack is to control the grain size and morphology. Small grain size causes a high capillary pressure (p_c) of the liquid, whose value is obtained from [37]:

$$p_c = \frac{4\gamma}{h} \tag{1.7}$$

where γ and h are the surface tension and the thickness of the liquid between the equiaxed grains, respectively. Grain refinement causes the thickness of the liquid film on the grain boundaries to decrease, which subsequently increases the capillary pressure, which inhibits the hot crack propagation. The equiaxed grains are present at the centerline of the weld to reduce SC in contrast to the columnar grains [52]. This is because lower strain is concentrated in the equiaxed grains. Furthermore, in the fine equiaxed grains, higher grain boundaries exist, which permits the liquid to have a higher permeability through the grain boundary, causing higher deformation to be tolerated [52]. These two factors, the width of the equiaxed zone in the middle of weld (w_{eqx}) and the grain size of this region (d_{eqx}), are unified in a parameter named the width-specific number of grain boundaries (N_{GB}), which is obtained from [53]:

$$N_{GB} = \frac{w_{eqx}}{d_{eqx}} + 1 \tag{1.8}$$

Grain refinement in the welding zone can be obtained by addition of Ti, which promotes the nucleation and increases the width of equiaxed zone at the center by

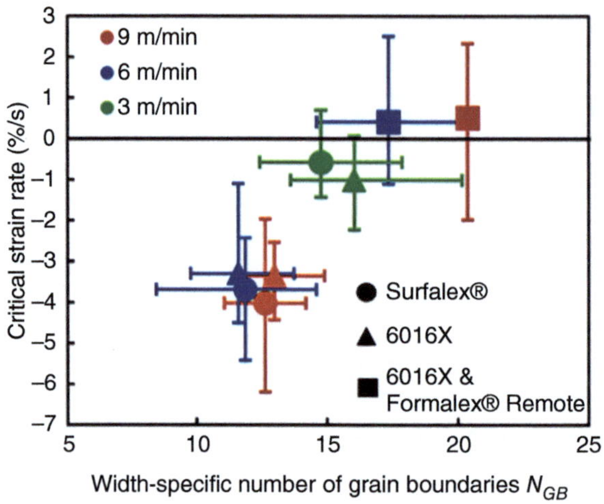

Figure 1.18 Dependency of critical strain rates of the weld with the width-specific number. Source: Ref. [53]/Sage Publications/CC BY 4.0.

restricting the growth during solidification, causing a transition from columnar to equiaxed growth [53]. Addition of Ti is more effective than controlling the welding parameters to change w_{eqx} and the grain size [53]. Figure 1.18 shows how the critical strain rate is increased when N_{GB} increases with the addition of Ti into the weld metal.

However, lowering the grain size below a critical size increases the susceptibility to hot cracking due to decreased permeability [54]. Thus, an optimum grain size exists that helps to mitigate SC. Figure 1.19 schematically presents how the grain size affects the SC sensitivity.

A criterion to figure out whether a composition of Al is not susceptible to cracking is when the following quantity is lower than a critical value [55]:

$$\left| \frac{dT}{d(\sqrt{f_S})} \right|_{\text{at } (\sqrt{f_{SS}}=0.95)} \tag{1.9}$$

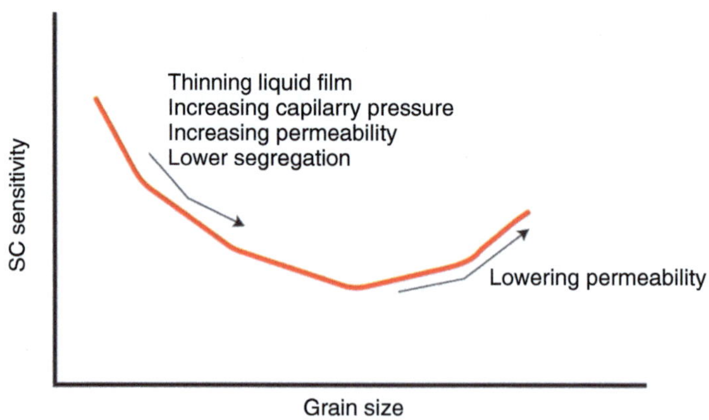

Figure 1.19 Schematic curve of the effect of grain size on SC sensitivity.

where T is the temperature, f_S is the solid fraction, and f_{SS} is the solid fraction at which the inter-dendritic separation occurs (about 0.9). The critical value is calculated from [55]:

$$\left| \frac{dT}{d(\sqrt{f_S})} \right| = \frac{2(1-k)(T_m - T_E)\sqrt{f_S}}{1 - f_S} \tag{1.10}$$

where k is the equilibrium segregation coefficient, T_E is the eutectic temperature, and T_m is the melting temperature. $\frac{dT}{d(\sqrt{f_S})}$ can be obtained using commercial thermodynamic software by plotting the temperature as a function of the solid fraction. In this way, it is possible to predict whether a composition is susceptible to SC or not. A high value of this index near $\sqrt{f_S} = 1$ indicates that the neighboring dendrites are not bridged well to resist cracking by slowing the lateral growth rate of the grains, where R is the characteristic value of the grain (Figure 1.20) [56].

An example is presented for Al—Si alloy, whose binary phase diagram is shown in Figure 1.21a. According to this figure, the following values are obtained:

$$(1-k)(T_m - T_E) = \Delta T_{max} = 75 \tag{1.11}$$

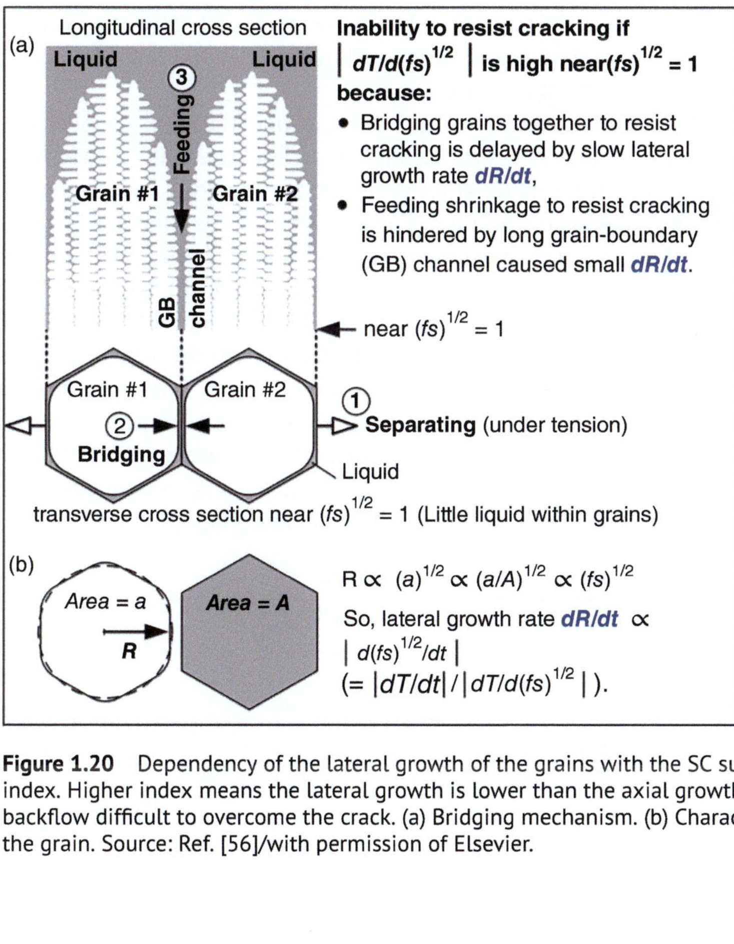

Figure 1.20 Dependency of the lateral growth of the grains with the SC susceptibility index. Higher index means the lateral growth is lower than the axial growth, making the backflow difficult to overcome the crack. (a) Bridging mechanism. (b) Characteristic value of the grain. Source: Ref. [56]/with permission of Elsevier.

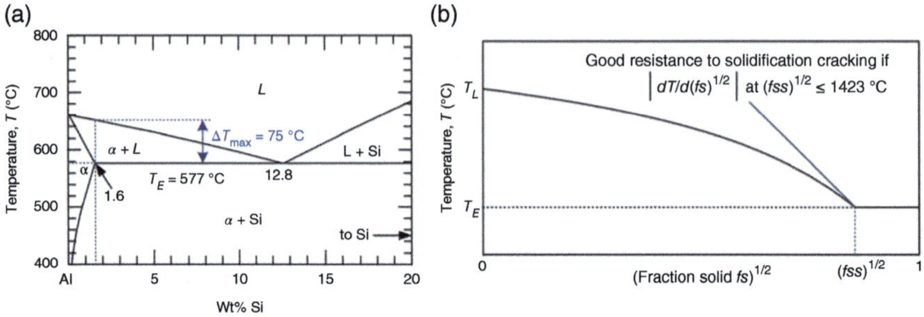

Figure 1.21 (a) Al/Si binary phase diagram. (b) Plot used to predict the SC by knowing the critical value. Source: Ref. [55]/Sage Publications.

$$k = \frac{1.6}{12.8} = 0.125 \tag{1.12}$$

Therefore:

$$\left| \frac{dT}{d(\sqrt{f_S})} \right| = \frac{2\Delta T_{\max}\sqrt{f_S}}{1 - f_S} = \frac{150\sqrt{0.9}}{1 - 0.9} = 1423\,°\mathrm{C} \tag{1.13}$$

The quantity $\left| \frac{dT}{d(\sqrt{f_S})} \right|$ is used as SC susceptibility index. The plot of temperature versus the square root of the solid fraction is shown in Figure 1.21b. This plot changes when using various filler materials and various dilutions. This plot helps to predict the SC sensitivity by using the above equation.

The choice of filler material is based on the flow behavior of the melt, the mechanical strength of the weld, and the hot cracking tendency of the base material [19]. This equation helps to select a filler material for a susceptible Al alloy to overcome the SC during welding by knowing the dilution. The limitation of this approach is that it is based on binary phase diagram, meaning that it neglects other alloying elements not present in the binary phase diagram. For ternary systems, similar curves like Figure 1.21b can be obtained for various amounts of one element while the other element is held constant [57]. It also does not take into account the diffusion in the solid state by which the alloying elements at the solidification front may deviate from the equilibrium state far from the phase diagram. The back diffusion from the inter-dendritic liquid to the solid dendrites decreases the SC susceptibility [57]. By considering the back diffusion, the solid fraction is modified, and consequently, the SC susceptibility index is modified according to the following equation [58]:

$$\left| \frac{dT}{d(\sqrt{f_S})} \right| = \frac{2(1-k)(T_m - T)\sqrt{f_S}}{1-(1-2\alpha' k)f_S} = \frac{2(1-k)(-m_L)C_o\sqrt{f_S}}{1-(1-2\alpha' k)f_S}[1-(1-2\alpha' k)f_S]^{\frac{k-1}{1-2\alpha' k}} \tag{1.14}$$

where α' is obtained from:

$$\alpha' = \alpha\left[1 - \exp\left(-\frac{1}{\alpha}\right)\right] - \frac{1}{2}\exp\left(-\frac{1}{2\alpha}\right) \tag{1.15}$$

$$\alpha = \frac{4D_s t_f}{\lambda_2^2} \tag{1.16}$$

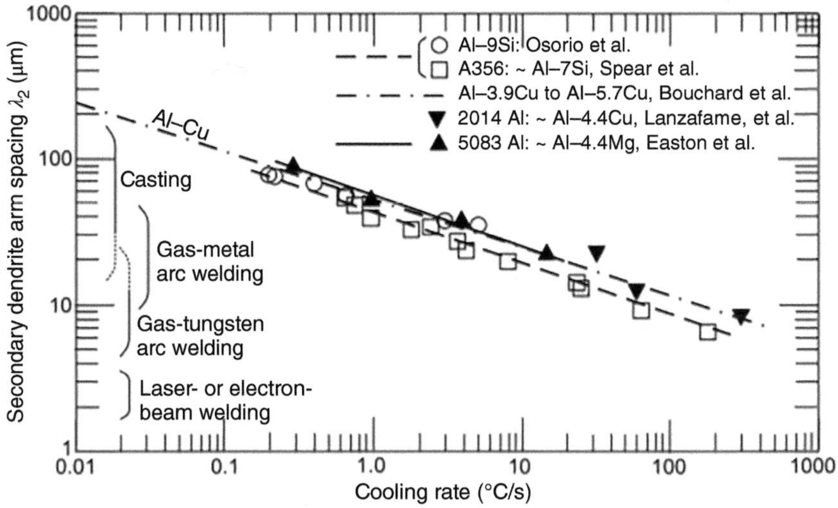

Figure 1.22 Secondary dendrite arm spacing of some Al alloys as a function of cooling rate. Source: Ref. [58]/with permission of Elsevier.

D_s is the diffusion coefficient of solute in solid and t_f is the local solidification time. In the extreme case where $\alpha' = \alpha = 0$, the simple criterion based on the Scheil equation is obtained where no back diffusion occurs $\left(\left|\frac{dT}{d(\sqrt{f_S})}\right| = \frac{2(1-k)(T_m-T_E)\sqrt{f_S}}{1-f_S} \text{ Eq. (1.10)}\right)$. The Scheil equation assumes that there is an equilibrium at the interface of liquid and solid, there is a complete diffusion in liquid, and there is no diffusion in solid. In another extreme case where $\alpha = \infty$, a complete solid-state diffusion occurs and lever-arm rule for an equilibrium state solidification is held. The secondary dendrite arm spacing λ_2 is a determinant factor, which itself is dependent on the cooling rate. The higher the cooling rate, the lower the secondary arm spacing. The range of this quantity is shown for various welding processes as well as casting in Figure 1.22. The back diffusion can more effectively reduce the SC when the cooling rate is low. The reason that back diffusion reduces the SC is the consumption of the alloying element in the liquid inside the inter-dendritic space [56].

Another modification to the SC susceptibility index based on the Scheil equation is developed for various solidification rates. The basis of this approach is to consider the effect of solidification rate on the T–f_S curve [59]. The downward shift of this curve results from the decrease in the eutectic temperature due to fast solidification rate (Figure 1.23a). The new SC susceptibility indexes derived from the new T–f_S curves for various processes with their specific solidification rates, are shown in Figure 1.23b. The maximum index increases with an increasing solidification rate. Also, the composition at which the maximum sensitivity occurs increases. Using the previous approach based on back diffusion, the laser welding with smaller α' value results in a smaller composition at which the maximum susceptibility exists, which does not align with the experiments.

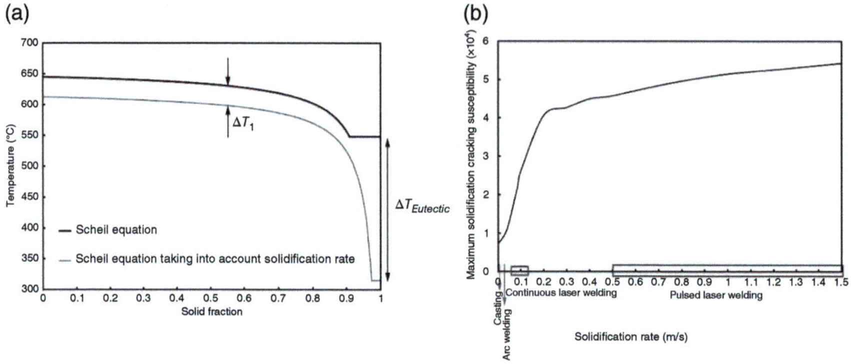

Figure 1.23 (a and b) $T-f_S$ curves based on the Scheil equation taking into account the solidification rate. Source: Ref. [59]/Springer Nature.

This approach is used to compare the weldability of a composition system at various solidification rates. To compare various composition systems (Al binary systems such as Al—Cu, Al—Mg, and Al—Zn), the back diffusion at eutectic temperature through the parameter αk is included in the SC susceptibility index. In this approach, the maximum value of $\left|\frac{dT}{d(\sqrt{f_S})}\right|$ is considered as an index for SC susceptibility. This approach justifies well why Al—Mg alloys, despite having a wide solidification temperature range, show a lower crack susceptibility due to their high value of k at eutectic temperature [56]. Figure 1.24a and b show the temperature versus the square root of the solid fraction when $\alpha = 0.05$ for Al—Cu and Al—Mg alloys, respectively. The corresponding crack susceptibility indexes versus the alloy content for various values of α are shown in Figure 1.24b and c. A significant decrease in the maximum index is observed for Al—Mg alloy at high values of α, which is due to the high value of k.

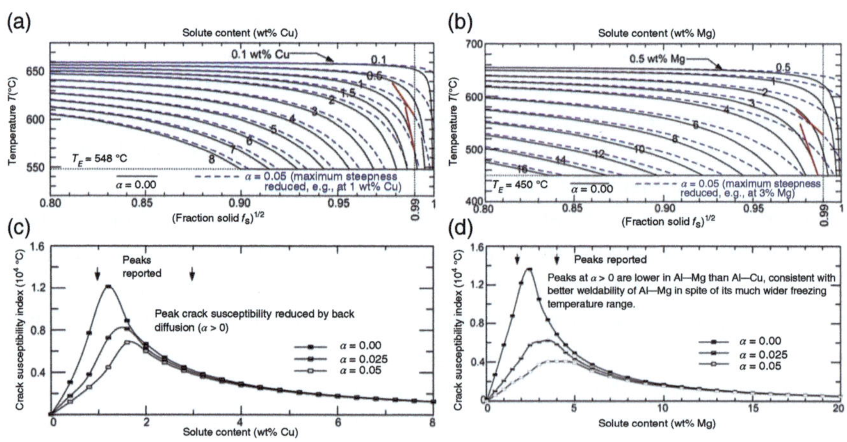

Figure 1.24 Temperature versus the square root of the solid fraction for (a) Al–Cu and (b) Al–Mg alloys. The corresponding crack susceptibility indexes versus the alloy content for various values of α for (c) Al–Cu and (d) Al–Mg alloys. Source: Ref. [56]/with permission of Elsevier.

In some processes like laser beam welding, the vaporization leads to a change of alloying elements [60], which is not considered in the equations of SC susceptibility index. For instance, the burning of Mg element during laser beam welding of 6xxx alloys leads to a decrease in the strength of the weld, and hence, 5xxx filler wires are used to compensate for the Mg content [61]. This needs to be taken into account when using these indexes.

A variant of hot cracking that locally occurs close to the FZ in the PMZ is liquation cracking, which is caused by melting the constituents of low melting temperatures such as eutectics or inclusions. This type of crack is also more likely for precipitation-hardened alloys, especially when there is no time during heating to dissolve these precipitates. The liquation occurs close to the FZ in the grain boundaries. Due to a large extent of the semisolid region and high thermal conductivity of Al, the region of PMZ is relatively large during welding Al alloys. A simple criterion to predict the occurrence of liquation cracking in PMZ is explained by the solid fraction in both PMZ and FZ adjacent to the PMZ [43]. The contraction stress imposed by the FZ on the PMZ can cause liquation cracking when the solid fraction in FZ is higher than the one in the PMZ, as shown in Figure 1.25b. In this state, the strength of the PMZ is not enough to resist the contraction caused by solidification in the FZ. When the solid fraction in PMZ is higher than in FZ, it can resist the contraction and no liquation cracking occurs (Figure 1.25b). Seemingly, the supply of the liquid from the FZ to the cracks in PMZ heals the crack in this region and prevents the liquation cracking. The occurrence of liquation cracking also depends on the microstructure and the temper of the base material. For instance, AA6061-T6 is more susceptible to liquation cracking than AA6061-T4, as the former has a coarser grain size with larger amounts of Si-rich eutectic compounds at the grain boundaries [62].

The composition of the Al alloy determines the thermal conductivity and hence the solidification rate [35]. This in turn determines the strain rate and hence the susceptibility to SC. So, the effect of chemical composition on SC is multifactorial, whose effect can be understood in more detail by simulation.

1.3 Thermal Cycle and Its Effect on the Weld Zone

Precipitation hardening alloys of Al undergo a thermal cycle in the HAZ, which can cause over-aging through which semi-coherent θ' phase converts to more stable θ. This causes a softening in the HAZ of the welds, which deteriorates the integrity of the structure. The microhardness test is a useful method to measure the mechanical properties in various regions of the welds by giving a hardness profile resolving different regions. A schematic of the microhardness distribution in HAZ with respect to the distance from the FZ boundary is shown in Figure 1.26. Adjacent to the FZ boundary, the hardness is lowest due to the solution heat treatment caused by the heat of the welding process (zone 1). In zone 3, the precipitates change to more stable θ and the percentage of this change increases with an increase in the pick temperature as it approaches the FZ boundary. Right adjacent to the FZ boundary, also named partially melted zone, the hardness can be recovered partially

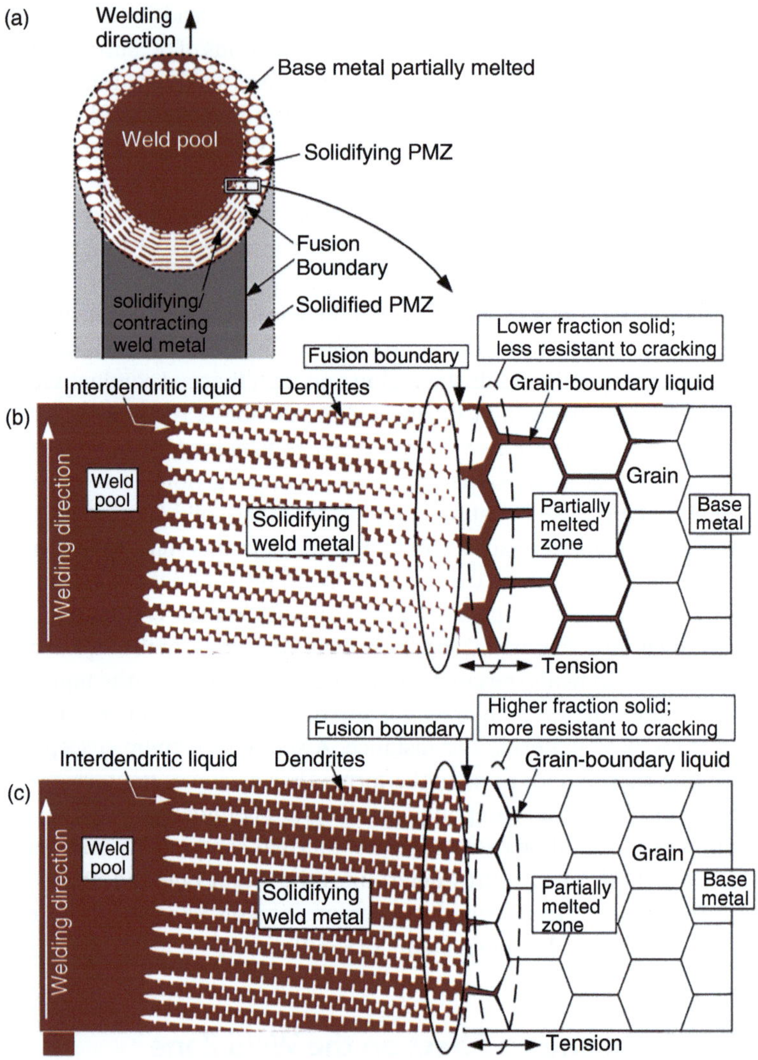

Figure 1.25 (a) Top view of the weld metal. (b) PMZ susceptible to liquation cracking. (c) PMZ resistant to liquation cracking. Source: Ref. [43]/Springer Nature.

by natural aging after welding (region *B* in Figure 1.26) [63]. Through an artificial aging after welding, the hardness of this region can be recovered completely (region *A* in Figure 1.26). Farther from this region toward the base material is a region that is partially solutioned and mainly in the over-aged state. Through artificial aging, the hardness can be partially recovered leaving this region to have the lowest hardness (region 2 in Figure 1.26).

To fully recover the original hardness, a PWHT consisting of a solutioning heat treatment and subsequent aging is needed [64]. In this way, the stable phases formed in the over-aged regions dissolve during solutioning and change to fine and uniformly dispersed semi-coherent precipitates during aging.

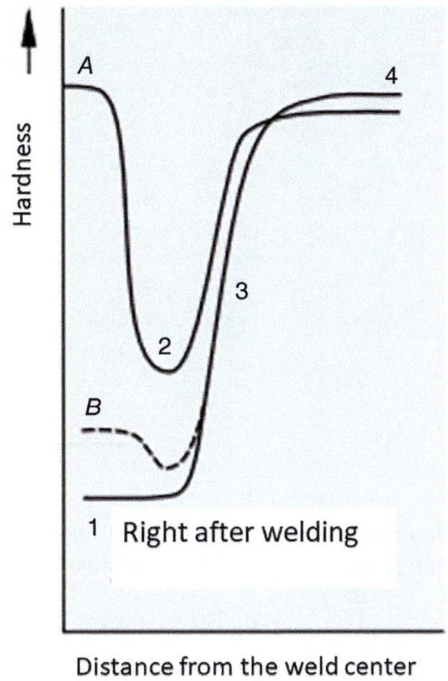

Figure 1.26 Schematic of hardness distribution in the HAZ of welded age-hardened aluminum alloy (curve 1). Source: Ref. [2]/Springer Nature.

Figure 1.27 shows a schematic of decrease in hardness in the weld zone and HAZ of a HTA welded by the gas metal arc welding (GMAW) process. The microstructure and the consequent mechanical properties (hardness) are recovered through solutioning and aging, as can be observed in Figure 1.27.

One strategy to overcome the softening in the HAZ is using the welding processes with a low heat input, such as laser welding, friction stir welding (FSW), and cold metal transfer (CMT) [65]. These processes are discussed in the following chapters.

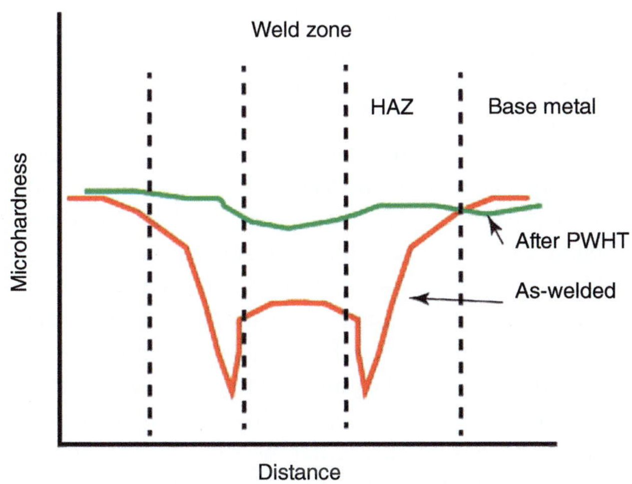

Figure 1.27 Schematic of microhardness profile of a HTA welded by GMAW. Microhardness profile after PWHT.

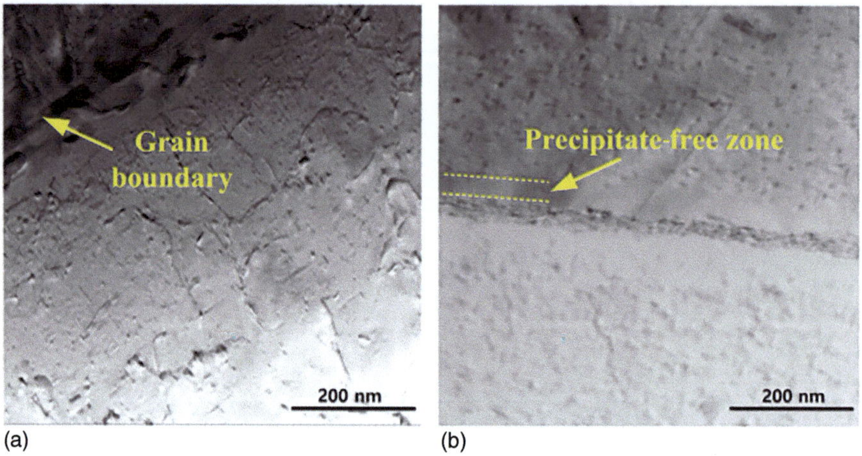

Figure 1.28 (a) FZ of an AA6061 alloy after aging. (b) The FZ after solution treatment and aging. Source: Ref. [67]/with permission of Elsevier.

The response of the FZ and HAZ is not similar to heat treatment. The FZ of Al welds is usually in the solution-treated state while the HAZ can be in various conditions, as mentioned. Due to the segregation of the alloying elements in the liquid phase at the final stage of solidification in the FZ, the solid may have insufficient concentration of elements, which makes it hard to respond to aging heat treatment after welding. An aging at proper time and temperature after welding may recover the hardness in the FZ depending on the degree of segregation [65]. Dislocations may also be present in the FZ of the welds. In the transmission electron microscopy (TEM) image of the FZ of a 6061-T6 Al alloy, no precipitate is observed and some dislocations are discernible. By aging after welding, the dislocations disappear and the precipitates appear [66].

As the final inter-dendritic liquid is rich in alloying elements, depending on the alloy type, some crystal phases form at the grain boundaries which are different from the phases that form during aging of the base material. These inter-granular crystals do not contribute to the strength of the weld, and instead, they reduce the degree of supersaturation inside the grains. Figure 1.28a shows the FZ of an AA6061 alloy welded by resistance welding, which is subsequently aged. The crystals in the grain boundary are discernible. Inside the grain, only a partial precipitation is observed, which is due to the insufficient amount of the alloying elements. When the FZ is solution-treated after welding, the interior of the grains is covered by massive amounts of precipitates during age hardening. A narrow region that corresponds to precipitation-free zone is observed around the grain boundaries (Figure 1.28) [67].

Talking about heat-treatable Al alloys, in solid-state welding processes such as FSW, where no solidification occurs, a nugget zone exists where the precipitates are likely to dissolve due to the heat of the friction. The hardness profile after FSW is dependent on the initial heat treatment state of the Al alloy [68]. When the alloy is in O-temper condition, the nugget zone has the highest hardness due to grain refinement (Figure 1.29). When the Al alloy is precipitate-hardened, for instance T6

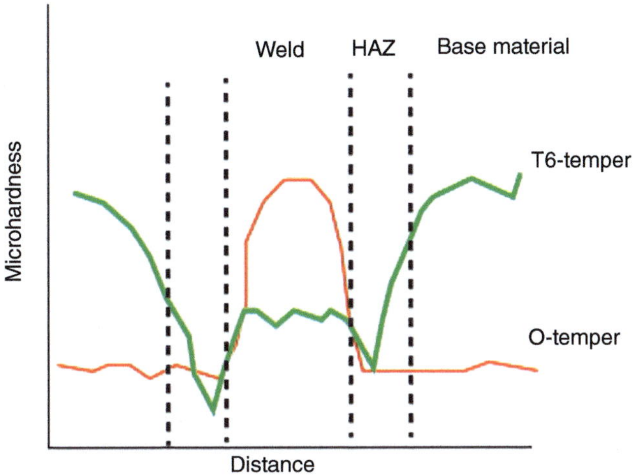

Figure 1.29 Schematic of hardness profile after FSW of O-temper and T6-Temper Al alloy.

condition, the nugget zone has the lowest hardness (Figure 1.29) due to dissolution of the precipitates and elimination of the dislocations. This is despite the finer grains in the nugget zone, because the contribution of the other factors is predominant [68]. So, the main challenge exists for FSW of those Al alloys that are hardened by precipitation. To recover the strength, a fully solutioning and aging treatment is needed in these joints [66]. This kind of heat treatment is not feasible for large structures. Moreover, a large distortion is induced during this heat treatment. Another issue caused by solutioning is the abnormal grain growth in the nugget zone. After FSW, the grains in the nugget zone are refined, which contributes to the weld strength. By subsequent solutioning and abnormal grain growth, a decrease in the strength as well as in the ductility is observed [21]. This decrease is compensated through precipitation by the following aging [69]. The elimination of the solutioning after FSW and only doing artificial aging, the decrease in strength due to grain growth and lower dislocation density is not observed anymore and hence a higher strength is obtained [70].

A detailed characterization of the precipitates and the dislocations needs high-resolution techniques such as TEM. The statistical reliability of this technique depends on the selection basis of the area of the study, as this technique yields useful information in a very small area below 10 μm. Microhardness profile data along with optical microscopy images can be used as bases for selection of the areas to be studied by TEM. The areas with similar hardness and morphology can be represented by choosing a random point for TEM analysis. Useful statistical data of the grains' crystallographic orientation and morphology can be obtained using electron back-scattered diffraction. The morphology of the grains of the weld zone depends on the temperature gradient (G) and the solidification speed (R) (Figure 1.30). The cooling rate (GR) determines the scale of the microstructure and G/R determines the morphology [71]. The extent of each morphology in the FZ depends on the welding process and welding parameters. These are addressed

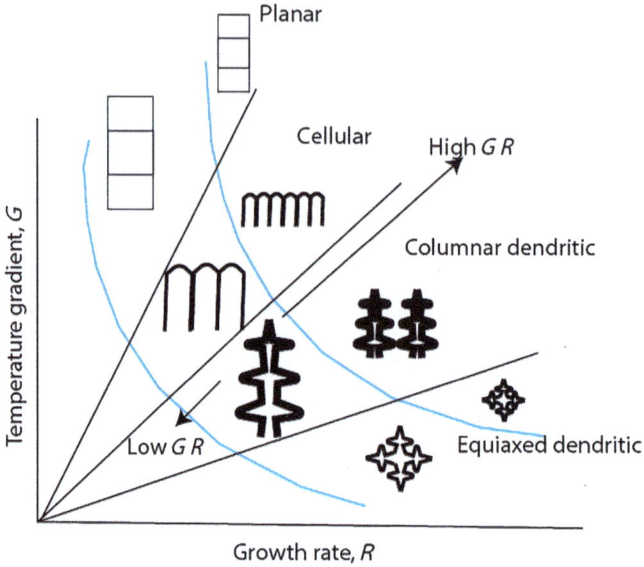

Figure 1.30 Schematic showing the effect of the temperature gradient (*G*) and the solidification speed (*R*) on the morphology and size of the grains in FZ.

in detail in the following chapters for arc welding processes, laser beam welding, resistance welding, and solid-state welding.

The mechanical properties of the weld depend on the microstructure. The microstructure of the FZ is cast, which consists of dendrites or columnar grains. Equiaxed grains may also appear in the weld zone, which is favored over the columnar or dendritic grains. It is shown that the equiaxed grains result in a higher ductility of the weld without any influence on the tensile strength [72]. The fine equiaxed structure provides more grain boundaries, which act as dislocation barriers, causing an increase in the yield strength (σ_y) according to the Hall–Petch equation [73].

$$\sigma_y = \sigma_0 + k.d^{-0.5} \tag{1.17}$$

where σ_0 is the opposite shear stress to the movement of dislocation inside the grain, k is a material constant representing the degree of difficulty to produce grain dislocation, and d is the grain diameter. The grain structure may also affect the mechanical properties through other mechanisms. For example, segregation of the alloying elements that cause hot cracking can also affect the failure mechanism of the joint in loading. Segregation is influenced by the grain structure, as discussed before. Along with the grain size and morphology, other mechanisms such as precipitation hardening, solution hardening, and strain hardening need to be considered in analyzing the mechanical properties of the joint.

In addition to thermal history, the nucleation particles affect the mechanical properties through modification of the solidification structure [74]. The metallographic images of the failed tensile specimens from the weld cross sections

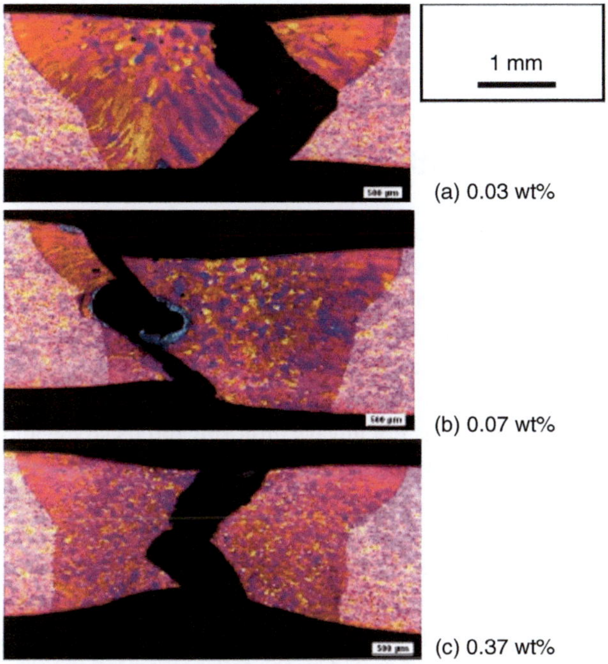

Figure 1.31 Metallographic images of the failed tensile specimens from the weld cross sections of three AA6082Al alloys with three different Ti content welded by laser welding [74]/DVS Media GmbH.

of three AA6082Al alloys with three different Ti content welded by laser welding are shown in Figure 1.31. At low Ti content, the FZ boundary consists of coarse columnar grains, where their boundary act as the fracture site during tensile testing. By 0.07% Ti addition, the fracture occurs along the boundary between the columnar and equiaxed grains. By 0.37% Ti, a full fine-equiaxed grain structure is developed, which promotes higher plastic deformation. The fracture occurs in the middle of the weld zone. The yield strength and ultimate tensile strength (UTS) of three specimens differ slightly, but the elongation of the weld to the fracture is highest in the specimen with the highest amount of Ti. The underlying mechanism by which the failure of the weld is influenced is the segregation and hot cracking caused by the grain structure. The degree of segregation is higher in coarse and columnar grains than the fine and equiaxed ones, as mentioned before. This explains the fracture locations observed in Figure 1.31. The mechanism of grain modification by adding nucleation particles is discussed in more detail in Chapter 2.

The thermal cycle not only influences the microstructure and mechanical properties, but also the distortion and residual stresses during welding. The distortion of Al alloys during welding is higher than steel due to its higher heat conductivity and higher thermal expansion coefficient. There are some basic laws that are practical for avoiding distortion during welding. For fillet welds, two simultaneous welding on both sides reduces the angular distortion. The seam welds, which cause the highest stiffness on the structure, should be performed last. The butt welds should be done before the filler welds [75].

1.4 Welding of Aluminum to Other Metals

A high demand exists for joining of Al to other metals such as steel, magnesium, and titanium, mainly for structural reasons. Welding of Al to copper is also gaining attention for electrical purposes, and hence, Al usage in the energy sector is increasing. The main challenge of welding Al to these materials is the formation of thick intermetallic compounds (IMCs), which are brittle and degrade the mechanical properties of the joint [76]. Usually, the interface is composed of one or more IMC layers at the interface. However, this layer is not always uniform, and its shape and morphology depend on the process and its parameters. The mechanical properties of the joint are highly influenced by the thickness and distribution of IMC as well as the direction of the applied load with respect to the interface [77, 78]. Dependency of the joint strength to the IMC thickness is depicted in Figure 1.32, which shows that while IMCs formation is necessary to establish a joint, its growth needs to be controlled to maintain its thickness in an optimum range.

The thickness of the IMC layer (d_{IMC}) is estimated by the following relationship

$$d_{IMC} = Dt^n \tag{1.18}$$

where D is the diffusion coefficient, t is time, and n is a constant whose value depends on the controlling factor of the growth, diffusion or reaction [79]. In dissimilar joints, the microstructures of the parent materials are not as decisive as the IMC layer, as this layer is the weakest point that controls the mechanical behavior. The mechanisms of IMC formation in fusion welding and solid-state welding processes are discussed in the following chapters.

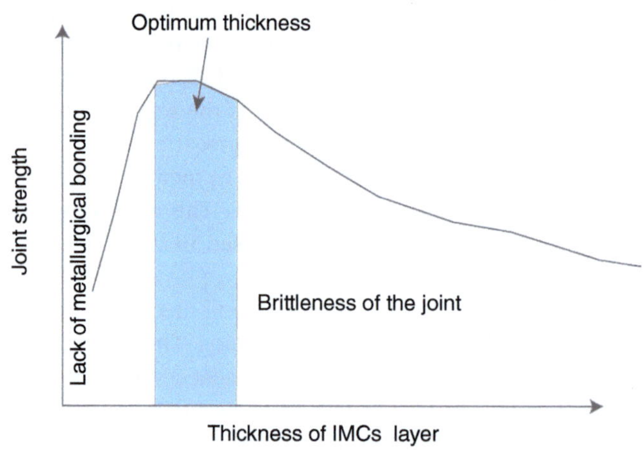

Figure 1.32 Dependency of the joint strength to the IMC thickness.

References

1 Murphy, A., McCune, W., Quinn, D. et al. (2007). The characterisation of friction stir welding process effects on stiffened panel buckling performance. *Thin-Walled Structures* 45 (3): 339–351.

2 Schulze, G. (2004). *Die Metallurgie des Schweißens: Eisenwerkstoffe—Nichteisenmetallische Werkstoffe*. Springer.

3 Dorn, L. (1998). Schweißverhalten von Aluminium und seinen Legierungen. *Materials Science and Engineering Technology* 29 (8): 412–423.

4 Zhang, P., Jia, Z., Yu, Z. et al. (2021). A review on the effect of laser pulse shaping on the microstructure and hot cracking behavior in the welding of alloys. *Optics Laser Technology* 140: 107094.

5 Srikanth, T., Surendran, S., Balaganesan, G. et al. (2017). Response of CMT welded aluminum AA5086-H111 to AA6061-T6 plate with AA4043 filler for ballistic. *Procedia Engineering* 194: 522–528.

6 Gould, J. (2012). Joining aluminum sheet in the automotive industry – a 30 year history. *Welding Journal* 91 (1): 23–34.

7 Sun, T., Franciosa, P., Sokolov, M. et al. (2020). Challenges and opportunities in laser welding of 6xxx high strength aluminium extrusions in automotive battery tray construction. *Procedia CIRP* 94: 565–570.

8 Abioye, T.E., Olanipekun, E., and Anasyida, A.S. (2022). Prediction of the hardness of pulverized glass waste-reinforced aluminium alloy 6061-T6 friction stir welded joint. *International Journal of Engineering Research in Africa* 62: 31–42.

9 Merino, J., Patzelt, A., Steinacher, A. et al. (2017). Ariane 6-Tanks and structures for the new european launcher. *Deutscher Luft- und Raumfahrtkongress*, München. Deutsche Gesellschaft für Luft-und Raumfahrt-Lilienthal-Oberth eV.

10 Cai, C., Shuang, H., Chen, H. et al. (2019). The influences of Ar—He shielding gas mixture on welding characteristics of fiber laser-MIG hybrid welding of aluminum alloy. *Optics & Laser Technology* 113: 37–45.

11 Lee, H.-K., Park, S.-h., and Kang, C.-Y. (2015). Effect of plasma current on surface defects of plasma-MIG welding in cryogenic aluminum alloys. *Journal of Materials Processing Technology* 223: 203–215.

12 Kashaev, N., Ventzke, V., and Çam, G. (2018). Prospects of laser beam welding and friction stir welding processes for aluminum airframe structural applications. *Journal of Manufacturing Processes* 36: 571–600.

13 Hobbacher, A. (2016). *Recommendations for Fatigue Design of Welded Joints and Components*, vol. 47. Springer.

14 Hossain, I., Bhowmik, A., Pattanaik, A. et al. (2024). Computational investigation of plasma arc welding process for aluminium alloys. *Engineering Research Express* 6 (2): 025541.

15 Dobosy, Á., Gáspár, M., and Török, I. (2018). Resistance spot welding of 7075 aluminium alloy. *Proceedings of the Second Vehicle and Automotive Engineering 2: VAE2018*, Miskolc, Hungary. Springer.

16 Dilthey, U. (2006). *Schweißtechnische Fertigungsverfahren 2: Verhalten der Werkstoffe beim Schweißen*. Springer-Verlag.

17 Kovács, J. and Lukács, J. (2022). Investigation of the Possibility for Compensating the HAZ Softening of AA7075-T6. *Periodica Polytechnica Mechanical Engineering* 66 (3): 207–212.

18 Liu, T., Mu, Z., Hu, R. et al. (2019). Sinusoidal oscillating laser welding of 7075 aluminum alloy: hydrodynamics, porosity formation and optimization. *International Journal of Heat and Mass Transfer* 140: 346–358.

19 Enz, J. (2012). *Laserstrahlschweissen von hochfesten Aluminium-Lithium-Legierungen*. Springer.

20 Warner, T. Recently-developed aluminium solutions for aerospace applications. In Materials Science Forum. 2006. *Trans Tech Publications.*

21 Zhang, J., Feng, X.S., Gao, J.S. et al. (2018). Effects of welding parameters and post-heat treatment on mechanical properties of friction stir welded AA2195-T8 Al–Li alloy. *Journal of Materials Science & Technology* 34 (1): 219–227.

22 Avilov, V., Gumenyuk, A., Lammers, M. et al. (2012). PA position full penetration high power laser beam welding of up to 30 mm thick AlMg3 plates using electromagnetic weld pool support. *Science Technology of Welding Joining* 17 (2): 128–133.

23 Tang, G., Chen, H., Yang, X. et al. (2018). Effects of different welding process on the electronic temperature of plasma and weld shape during laser-MIG hybrid welding of A7N01P-T4 aluminum alloy. *Journal of Laser Applications* 30 (2): 022002.

24 Balanovskii, A.E. (2019). A new mechanism of interaction between a welding arc discharge of reverse-polarity direct current and an aluminum surface. *High Temperature* 57: 784–797.

25 Pellone, L., Inmake, G., Hong, K.-M. et al. (2019). Effects of interface gap and shielding gas on the quality of alloy AA6061 fiber laser lap weldings. *Journal of Materials Processing Technology* 268: 201–212.

26 Feng, J., Zhang, P., Yan, H. et al. (2023). Application of laser welding in electric vehicle battery manufacturing: a review. *Coatings* 13 (8): 1313.

27 Kang, Y., Zhan, X.H., and Liu, T. (2019). Effect of welding parameters on porosity distribution of dual laser beam bilateral synchronous welding in 2219 aluminum alloy T-joint. *Journal of Adhesion Science and Technology* 33 (23): 2595–2614.

28 Chen, Q., Ge, H., Yang, C. et al. (2017). Study on pores in ultrasonic-assisted TIG weld of aluminum alloy. *Metals* 7 (2): 53.

29 Ardika, R.D., Triyono, T., and Muhayat, N. (2021). A review porosity in aluminum welding. *Procedia Structural Integrity* 33: 171–180.

30 He, E., Liu, J., Lee, J. et al. (2018). Effect of porosities on tensile properties of laser-welded Al–Li alloy: an experimental and modelling study. *The International Journal of Advanced Manufacturing Technology* 95: 659–671.

31 Yan, S., Zhu, Z., Ma, C. et al. (2019). Porosity formation and its effect on the properties of hybrid laser welded Al alloy joints. *The International Journal of Advanced Manufacturing Technology* 104: 2645–2656.

32 Vyskoč, M., Sahul, M., and Sahul, M. (2018). Effect of shielding gas on the properties of AW 5083 aluminum alloy laser weld joints. *Journal of Materials Engineering and Performance* 27: 2993–3006.

33 Han, X., Yang, Z., Ma, Y. et al. (2020). Porosity distribution and mechanical response of laser-MIG hybrid butt welded 6082-T6 aluminum alloy joint. *Optics & Laser Technology* 132: 106511.

34 Chen, L., Wang, C., Xiong, L. et al. (2020). Microstructural, porosity and mechanical properties of lap joint laser welding for 5182 and 6061 dissimilar aluminum alloys under different place configurations. *Materials & Design* 191: 108625.

35 Miyagi, M., Kawahito, Y., Wang, H. et al. (2018). X-ray phase contrast observation of solidification and hot crack propagation in laser spot welding of aluminum alloy. *Optics Express* 26 (18): 22626–22636.

36 Rappaz, M., Drezet, J.-M., and Gremaud, M. (1999). A new hot-tearing criterion. *Metallurgical and Materials Transactions A* 30: 449–455.

37 Tang, Z. and Vollertsen, F. (2014). Influence of grain refinement on hot cracking in laser welding of aluminum. *Welding in the World, Le Soudage Dans Le Monde* 58: 355–366.

38 Wang, X.J., Lu, F., Wang, H.-P. et al. (2015). Experimental and numerical analysis of solidification cracking behaviour in fibre laser welding of 6013 aluminium alloy. *Science and Technology of Welding & Joining* 20 (1): 58–67.

39 Pumphrey, W.I. and Moore, D.C. (1948). A consideration of the nature of brittleness at temperatures above the solidus in castings and welds in aluminium alloys. *Journal of the Institute of Metals* 75: 235–256.

40 Bielenin, M. (2021). *Prozessstrategien zur Vermeidung von Heißrissen beim Schweißen von Aluminium mit pulsmodulierbaren Laserstrahlquellen*. BoD–Books on Demand.

41 Kubo, K. and Pehlke, R.D. (1985). Mathematical modeling of porosity formation in solidification. *Metallurgical and Materials Transactions B* 16: 359–366.

42 Kou, S. (2003). *Welding Metallurgy*, 223–225. John Wiley & Sons.

43 Kou, S., Firouzdor, V., and Haygood, I.W. (2011). Hot cracking in welds of aluminum and magnesium alloys. *Materials Science Engineering* 3–23.

44 Jennings, P.H. (1949). *Hot-shortness of Some High-purity Alloys in the Systems Aluminium–Copper–Silicon and Aluminium–Magnesium–Silicon*. Aluminium Development Association.

45 Coniglio, N. and Cross, C.E. (2009). Mechanisms for solidification crack initiation and growth in aluminum welding. *Metallurgical and Materials Transactions A* 40: 2718–2728.

46 Tian, Y., Robson, J.D., Riekehr, S. et al. (2016). Process optimization of dual-laser beam welding of advanced Al–Li alloys through hot cracking susceptibility modeling. *Metallurgical Materials Transactions A* 47: 3533–3544.

47 Hagenlocher, C., Weller, D., Weber, R. et al. (2019). Analytical description of the influence of the welding parameters on the hot cracking susceptibility of laser beam welds in aluminum alloys. *Metallurgical and Materials Transactions A* 50: 5174–5180.

48 Von Witzendorff, P., Kaierle, S., Suttmann, O. et al. (2015). In situ observation of solidification conditions in pulsed laser welding of AL6082 aluminum alloys to evaluate their impact on hot cracking susceptibility. *Metallurgical and Materials Transactions A* 46: 1678–1688.

49 Hagenlocher, C., Stritt, P., Weber, R. et al. (2018). Strain signatures associated to the formation of hot cracks during laser beam welding of aluminum alloys. *Optics and Lasers in Engineering* 100: 131–140.

50 Coniglio, N., Cross, C.E., Michael, T. et al. (2008). Defining a critical weld dilution to avoid solidification cracking in aluminum. *Welding Journal* 87 (8): 237s–247s.

51 Wei, H., Chen, J.S., Wang, H.-P. et al. (2016). Thermomechanical numerical analysis of hot cracking during laser welding of 6XXX aluminum alloys. *Journal of Laser Applications* 28 (2): 022405.

52 Norouzian, M., Elahi, M.A., and Plapper, P. (2023). A review: suppression of the solidification cracks in the laser welding process by controlling the grain structure and chemical compositions. *Journal of Advanced Joining Processes* 7 (9): 100139.

53 Hagenlocher, C., Weller, D., Weber, R. et al. (2019). Reduction of the hot cracking susceptibility of laser beam welds in AlMgSi alloys by increasing the number of grain boundaries. *Science and Technology of Welding & Joining* 24 (4): 313–319.

54 Easton, M.A., Grandfield, J., John, S., D. et al. (2006). The effect of grain refinement and cooling rate on the hot tearing of wrought aluminium alloys. *Materials Science Forum* 519–521: 1675–1680. Trans Tech Publications.

55 Soysal, T. (2021). A criterion to find crack-resistant aluminium alloys to avoid solidification cracking. *Science and Technology of Welding & Joining* 26 (2): 99–105.

56 Liu, J. and Kou, S. (2016). Crack susceptibility of binary aluminum alloys during solidification. *Acta Materialia* 110: 84–94.

57 Liu, J. and Kou, S. (2017). Susceptibility of ternary aluminum alloys to cracking during solidification. *Acta Materialia* 125: 513–523.

58 Liu, J. and Kou, S. (2015). Effect of diffusion on susceptibility to cracking during solidification. *Acta Materialia* 100: 359–368.

59 Malekshahi Beiranvand, Z., Ghaini, F.M., Naffakh-Moosavy, H. et al. (2019). An analytical model for prediction of solidification cracking susceptibility in aluminum alloys taking into account the effect of solidification rate. *Metallurgical Transactions A* 50: 2835–2846.

60 Holzer, M., Hofmann, K., Mann, V. et al. (2016). Change of hot cracking susceptibility in welding of high strength aluminum alloy AA 7075. *Physics Procedia* 83: 463–471.

61 Deng, A., Chen, H., Zhang, Y. et al. (2023). Prediction of the influence of welding metal composition on solidification cracking of laser welded aluminum alloy. *Materials Today Communications* 35: 105556.

62 Wan, Z., Meng, D., Zhao, Y. et al. (2022). Improvement in tensile properties of 2219-T8 aluminum alloy TIG welding joint by PMZ local properties and stress distribution. *Journal of Manufacturing Processes* 839: 142863.

63 Jula, M., Dehmolaei, R., and Ranjbar, K. (2023). Softening, hardening, and precipitation evolution of the AA6082-T651 heat-affected zone caused by thermal cycles during and after welding. *Metals and Materials International* 29 (12): 3664–3678.

64 Pérez, J.S., Ambriz, R.R., Curiel López, F.F. et al. (2016). Recovery of mechanical properties of a 6061-T6 aluminum weld by heat treatment after welding. *Metallurgical and Materials Transactions A* 47: 3412–3422.

65 Cheng, J., Song, G., Zhang, X. et al. (2021). Review of techniques for improvement of softening behavior of age-hardening aluminum alloy welded joints. *Materials (Basel)* 14 (19): 5804.

66 Jie, Y., Wang, G., Li, S.-k. et al. (2019). Effect of post-weld heat treatment on microstructure and mechanical properties of welded joints of 6061-T6 aluminum alloy. *Transactions of Nonferrous Metals Society of China* 29 (10): 2035–2046.

67 Zhang, Y., Li, H., Li, Z. et al. (2021). Quantitative analysis of the solute redistribution and precipitate of Al–Mg–Si series alloys resistance spot welds after post–weld heat treatment. *Journal of Materials Research and Technology* 15: 5906–5919.

68 Chen, Y., Liu, H., and Feng, J. (2006). Friction stir welding characteristics of different heat-treated-state 2219 aluminum alloy plates. *Materials Science and Engineering: A* 420 (1–2): 21–25.

69 Aydın, H., Bayram, A., and Durgun, I. (2010). The effect of post-weld heat treatment on the mechanical properties of 2024-T4 friction stir-welded joints. *Materials & Design* 31 (5): 2568–2577.

70 Elangovan, K. and Balasubramanian, V. (2008). Influences of post-weld heat treatment on tensile properties of friction stir-welded AA6061 aluminum alloy joints. *Materials Characterization* 59 (9): 1168–1177.

71 Geng, S., Jiang, P., Shao, X. et al. (2020). Heat transfer and fluid flow and their effects on the solidification microstructure in full-penetration laser welding of aluminum sheet. *Journal of Materials Science & Technology* 46: 50–63.

72 Wang, L., Gao, M., Zhang, C. et al. (2016). Effect of beam oscillating pattern on weld characterization of laser welding of AA6061-T6 aluminum alloy. *Materials & Design* 108: 707–717.

73 Saifudin, S., Muhayat, N., Surojo, E. et al. (2022). Mitigation of porosity and residual stress on car body aluminum alloy vibration welding: a systematic literature review. *Automotive Experiences* 5 (3): 477–493.

74 Tang, Z., Schempp, P., Seefeld, T. et al. (2011). *Kornfeinung beim WIG-und Laserstrahlschweißen von Aluminiumlegierungen*. DVS-Verlag.

75 Nölle, P. (2011). Verzugarmes Schweißen von Aluminiumwerkstoffen. *Lightweight Design* 4 (2): 58–63.

76 Beygi, R., Kazeminezhad, M., and Kokabi, A.H. (2014). Microstructural evolution and fracture behavior of friction-stir-welded Al—Cu laminated composites. *Metallurgical and Materials Transactions A* 45: 361–370.

77 Beygi, R., Talkhabi, A.A., Mehrizi, M.Z. et al. (2023). A novel lap-butt joint design for FSW of aluminum to steel in Tee-configuration: joining mechanism, intermetallic formation, and fracture behavior. *Metals* 13 (6): 1027.

78 Beygi, R., Carbas, R.J.C., Marques, E.A.S. et al. (2023). Mechanism of toughness enhancement of brittle fracture by intermittent η-intermetallic in Al/Cu joint made by FSW. *Materials Science and Engineering A* 145907.

79 Springer, H., Kostka, A., Payton, E. et al. (2011). On the formation and growth of intermetallic phases during interdiffusion between low-carbon steel and aluminum alloys. *Acta Materialia* 59 (4): 1586–1600.

2

Arc Welding Processes

2.1 Introduction

Effective usage of arc welding processes to join Al alloys needs a comprehensive understanding of the challenges found during fusion welding of Al alloys, such as hot cracking, oxide formation, porosity formation, and distortion [1]. This is beside the understanding of the arc welding techniques and their parameters. Three arc welding processes are discussed in this chapter: TIG or gas tungsten arc welding (GTAW), MIG or GMAW, and plasma arc welding (PAW). The arc welding processes are used not only for joining, but also for additive manufacturing (AM) processes based on arc welding. The physics of solidification and cooling in both welding and AM are similar with huge differences in their parameters resulting from differences in cooling rate. In AM the heat sinking occurs through the previously deposited layers, which themself are hot. This is a one-directional heat transfer contrary to the welding where the heat transfer occurs in two or three dimensions. So, the cooling rate in welding is higher than AM [2]. Arc welding processes are usually used to join thin Al sheets, though for thick sheets MIG welding is more suited than TIG. The use of arc welding process for welding thick Al sheets is challenging as thick Al sheets are more likely to produce pores during welding [3]. Also, the solidification rate is higher for thick Al sheets, which promotes the pore formation and hot cracking [4]. The welds made by arc welding and using filler materials possess a reinforcement. The geometry of the cap affects the strain distribution during loading due to the existence of the toes at the corners of the cap. Hence, in multi-pass welding the last pass can be performed by swinging to reduce this effect by enlarging the area of strain concentration during loading [5].

The arc welding processes radiate ultraviolet (UV), and therefore, safety matters need to be taken for welders exposed to these processes. The allowable exposure to UV radiation is provided in European Union (EU) Directive 2006/25/EC. The amount of UV radiation depends on the process, current, materials to be welded, and the filler material [6]. Before introducing the processes, some terms need to be defined: pulsed alternating current (AC) and variable polarity (VP). VP refers to waveforms with varied polarities of negative and positive while in AC the portions of both polarities as well as their magnitude are equal.

Welding Metallurgy of Aluminium Alloys: Design, Processes, and Simulations, First Edition.
Reza Beygi, Eduardo André de Sousa Marques, Ricardo João Camilo Carbas, and Lucas Filipe Martins da Silva.

2.2 Processes

Shielded metal arc welding (SMAW) with coated electrodes, which is performed manually, is rarely used for welding Al. The main reason is the absorption of the moisture by the coating, which is decomposed into hydrogen during arc welding and produces pores [7].

The tolerance of arc welding processes regarding the position of the heat source and the gap distances is higher than in the laser welding process. These tolerances for various welding techniques are shown in Figure 2.1. This figure shows the allowable tolerance for the position of the energy source as well as the tolerance limit for the gap distance of the adjoints. Among the three arc welding processes, plasma welding has the lowest tolerance [8].

2.2.1 TIG Welding

TIG welding is most suited for some Al alloys such as AA5083, which is a non-HTA. It attains its strength through solution hardening and strain hardening [9]. The main issue of Al welding is the oxide layer. In arc welding processes, this oxide layer is removed when the electrode is positive (Direct Current Electrode Positive [DCEP] or reverse polarity). In this state, the cathode spots formed on the surface of Al chaotically evaporate the oxide film and heat the Al layer to the melting point. Then, the cathode spots that are melted are fixed, and these liquid spots begin to enlarge in diameter [10]. In TIG welding when the electrode polarity is negative, a high amount of heat is produced on the surface of the workpiece. This is favorable for welding steel, but in the case of Al, this heat cannot melt away the oxide layer. This is

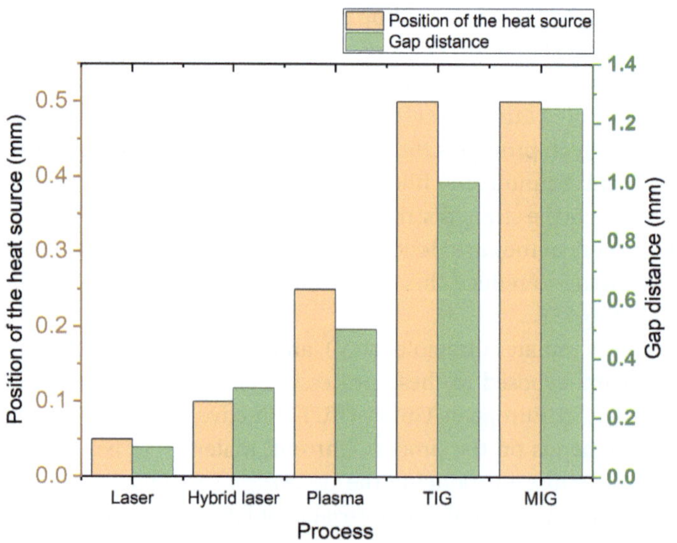

Figure 2.1 Position tolerance and weld-gap tolerance for various welding processes.

due to the high melting point of the oxide layer and the high heat conductivity of Al, which dissipates the heat [11]. So, welding of Al with non-consumable electrodes is performed in an alternating positive–negative pole, for example AC. By AC current, the portions of negative and positive poles are equal. When using rectangular AC current or direct current (DC) alternative pole current, the duration percentage of the poles can be adjusted. The percentage of the positive pole (electrode positive or EP%) affects the weld pool geometry and penetration when the current level is low. By increasing the EP% up to 20% the penetration decreases and beyond 20%, it increases, which is in contrast to the fact that the higher proportion of heat is produced at the positive pole. This is due to the field effect, which promotes the electron emission beyond the 20% EP [11]. The cold cathode Al emits electrode by this field emission, which leads to an increase in the energy input and hence a higher penetration and width. It also has a cleaning effect to remove the oxide layer by emission of the electrons from the oxide layer. The electrons are not emitted from the metal under the oxide layer, as the work required to emit the electrons from the oxide is half of that from the Al [12]. In fact, when it is said that the penetration in EP is lower, it is not related to the polarity itself, but to the low current capacity of the tungsten electrode, which is between 80 and 100 A [13]. Enlarging the diameter of the tungsten electrode to increase the current capacity is not a solution, as it causes the Lorentz force to decrease and the arc to be unstable. When the electrode is negative, there is no limitation for the current capacity of the tungsten electrode.

The main parameters of TIG welding are the current and the welding speed, considering that other parameters are selected correctly, such as the nozzle diameter, electrode diameter, flow rate of the protecting gas, and torch angle [14]. The best way to choose the correct parameters is a design of experiment to attain the optimum properties, such as maximum tensile strength [15, 16] or minimum susceptibility to hot cracking [17]. TIG welding machine acts in constant current mode, meaning that the voltage is automatically adjusted to reach the determined current. The oxide layer on the surface of Al acts as a resistance for the current to pass through the top surface. This causes a drop in voltage, and hence, higher voltage values of the machine are needed to compensate for this voltage drop. In VP-TIG welding, when the ratio of the EP increases, the needed voltage decreases, which is attributed to more efficient cleaning of the oxide layer in EP mode [18].

2.2.2 Plasma Arc Welding

PAW has developed from TIG welding where the arc between the non-consumable electrode and the workpiece is constricted by a nozzle and gas stream (Figure 2.2). This gas stream is ionized and contains a high number of charge carriers, causing the arc to become a plasma stream with a high energy density.

A comparison between the arcs of TIG and plasma is shown in Figure 2.3. PAW for Al is performed in AC, rectangular current waveform, and DC reverse polarity modes, the latter needs an intensive cooling system [19].

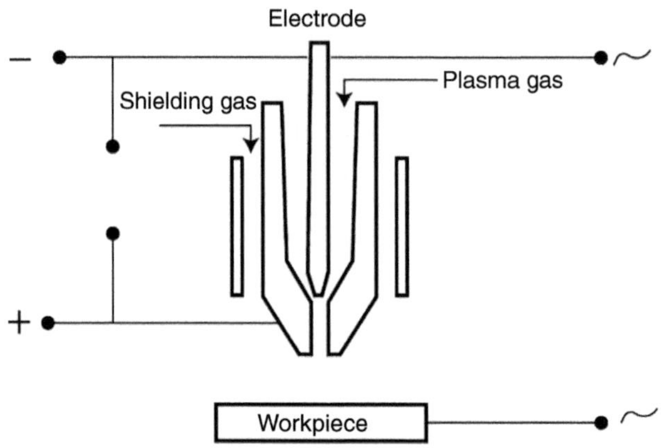

Figure 2.2 Schematic of plasma arc welding.

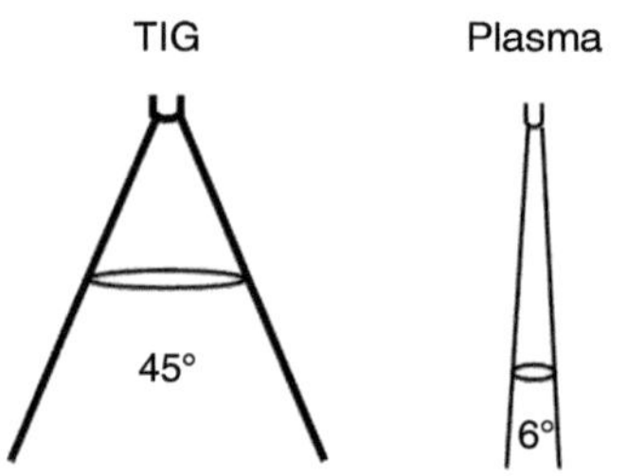

Figure 2.3 A comparison of arcs of TIG and PAW.

2.2.3 MIG Welding

MIG welding uses a consumable electrode and is performed at constant voltage. The penetration depth is higher when the electrode pole is positive and vice versa. When the electrode is negative, the temperature in the workpiece is reduced making it suitable for welding thin sheets. When the electrode is negative, the transfer of the droplets becomes unstable as it is performed only in globular form [20]. This is due to the formation of an asymmetrical magnetic field, which suspends the droplet on the wire tip. The usage of a VP, whose feature is shown in Figure 2.4, ensures a stable arc. Increasing the percentage of negative polarity ($\%EN$) decreases the penetration depth. This percentage is obtained from:

$$\%EN = \frac{T_n}{T_n + T_p + T_b} \tag{2.1}$$

When one droplet is detached in each pulse and the droplet diameter is equal to the diameter of the wire, a controlled transfer is produced, which is stable [21]. The average voltage curves versus average current during welding Al are shown in Figure 2.5. The increase in the average value of the current with increasing the wire feed speed (WFS) is obvious. When the arc voltage is high, the current is insensitive to the arc length and is constant. In this state, the mode of melt transfer is spray. Close to the short circuit conditions where the arc length is small, the correlation between the current and voltage is more complex [22]. The spray transfer exists when the

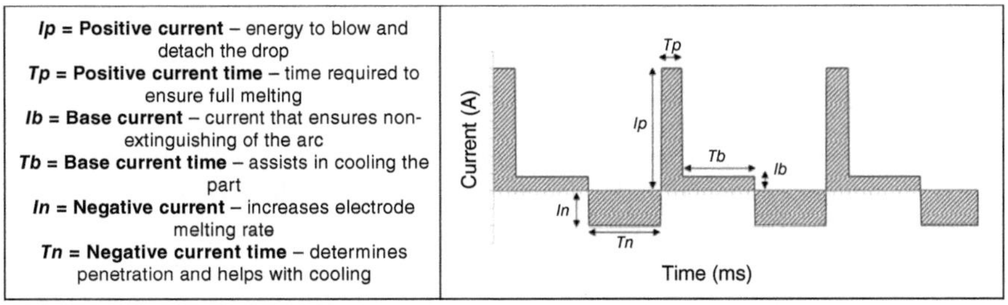

Figure 2.4 Schematic of variable polarity. Source: Ref. [20]/Seven Editora/CC BY 4.0.

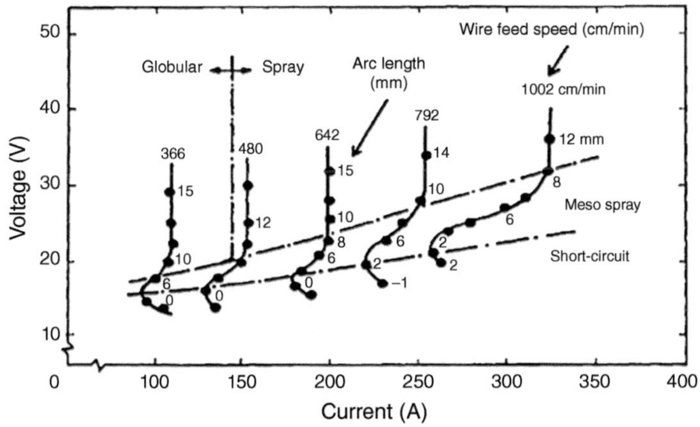

Figure 2.5 Behavior of arc during MIG welding of Al. Source: Ref. [22].

necking at the tip of the melted wire occurs and the droplet is detached before it contacts with the melt pool to form a short circuit transfer [23]. There is another metal transition, named globular transition, which occurs when the forces on the melted drop attached to the wire are higher than the forces tending to detach it. The magnetic pinch pressure (P_{Pinch}) is the main source of the force, which detaches the droplet from the wire and results from the self-magnetic field of the current. The main force that holds the droplet on the wire is the pressure resulting from the surface tension ($P_{Surface\ tension}$). The maximum values of these pressures are [24]:

$$P_{Surface\ tension} = \frac{2\gamma}{R_w} \tag{2.2}$$

$$P_{Pinch} = \frac{\mu_0 I^2}{4\pi^2 R_w^2} \tag{2.3}$$

where γ is the surface tension, R_w is the radius of the wire, I is the current, and μ_0 is a constant (permeability of free space). Equating these two pressures yields the current, below which the globular transfer occurs.

$$I = 2\pi \left(\frac{2\gamma R_w}{\mu_0} \right)^{0.5} \tag{2.4}$$

2.3 Modifications to the Processes

2.3.1 Negative-Pole and Positive-Pole TIG Welding

The main advantages of Direct Current Electrode Negative (DCEN) TIG welding are stable arc, increasing the depth penetration, utilizing lower currents, and higher speed of welding. The main issue when using negative pole for the electrode during TIG welding of Al alloy is the absence of the cleaning effect for removing the oxide layer. A newly developed variant of TIG welding, named Al-Gleichstromminuspolhandschweissen (AGL) welding process, has made it possible to perform the welding by negative electrode in which a gas mixture of He—Al with a higher content of He is used [8]. For the sheets with the thickness greater than 3 mm, this process can be used at a high welding speed, which reduces the distortion. Activated-flux TIG (A-TIG) welding in DCEN polarity can resolve this issue by removing the oxide layer through flux agents, the mechanism of which is described later.

When the electrode is positive, the heat is concentrated on the electrode, which causes it to melt and degrade fast. Furthermore, the arc is unstable and the penetration is low. The main advantage of this polarity is its cleaning effect on the oxide layer. Attempts have been made to perform TIG welding in DCEP. Designing new electrodes with effective cooling permits the use of higher currents up to 450 A. As the diameter of this electrode is high, a second gas flow around the electrode is used to increase the arc stability. This gas contains Ar with 300 ppm oxygen. This modified process can weld Al sheets with a thickness of 2–10 mm [13].

The needed time for electrode-positive for removing the oxide layer can be calculated automatically. When the surface of Al is very clean from oxide, the welding process continues in electrode-negative, making it possible to join thin Al sheets with a very low distortion. Bilateral synchronous TIG welding is another method used to reduce distortion during welding Al. Two sides of the joints are joined simultaneously by two TIG welding machines, which are synchronized with another machine in a way that the arcs do not influence each other. By this method, Al sheets with a thickness of 6 mm can be welded without any edge preparation with lower current [8]. Bilateral synchronous TIG welding reduces the distortion during welding when it is performed asymmetrically [25].

2.3.2 Keyhole TIG (K-TIG) Welding

K-TIG is a variant of TIG welding where a keyhole is formed by using high current values above 300 A. The arc pressure is very high, which penetrates the molten pool and makes an opening at the back of the weld. A system is needed to cool the tungsten electrode due to the high current [26]. A part of plasma that is formed inside the keyhole spurts out of the keyhole exit, which is named efflux (Figure 2.6). This process is used to weld Al with a thickness range of 6–12 mm. In comparison with laser welding, it has a higher tolerance to the weld gap, similar to what was mentioned earlier for conventional TIG.

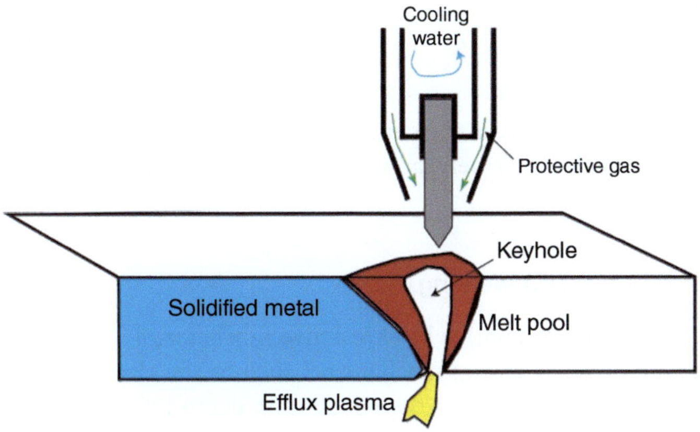

Figure 2.6 Schematic of K-TIG.

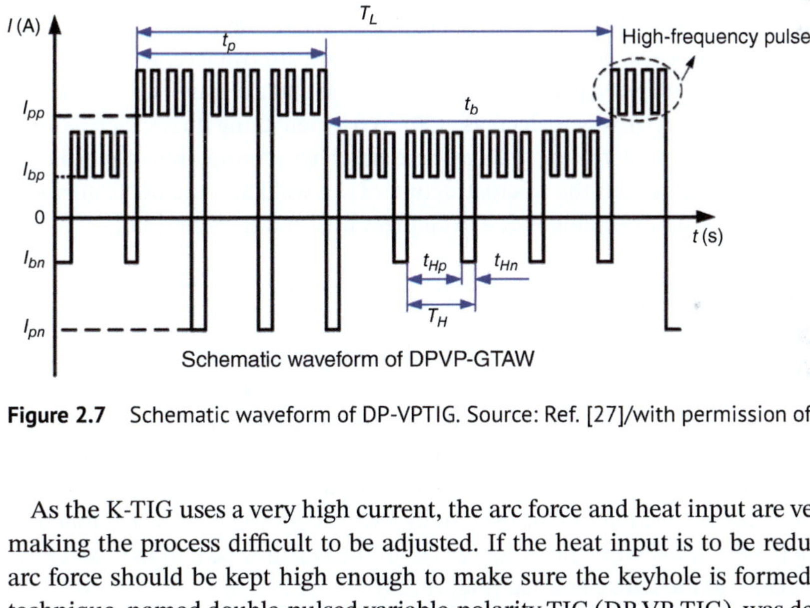

Figure 2.7 Schematic waveform of DP-VPTIG. Source: Ref. [27]/with permission of Elsevier.

As the K-TIG uses a very high current, the arc force and heat input are very high, making the process difficult to be adjusted. If the heat input is to be reduced, the arc force should be kept high enough to make sure the keyhole is formed. A new technique, named double-pulsed variable-polarity TIG (DP VP-TIG), was developed to make use of the advantages of both low-frequency and high-frequency pulses to solve this problem [27]. The schematic of the waveform of DP VP-TIG and its parameters are shown in Figure 2.7. In this process, a VP current was modulated into a low-frequency pulse (0.5–5 Hz) and a high-frequency pulse (20–100 kHz) superimposed on the positive polarity stage. I_{pp} helps to make the keyhole, while I_{bp} helps to reduce the heat input.

The dynamic of the keyhole formation in K-TIG differs from laser welding (refer to Chapter 3), as the acting forces and the mechanism of keyhole formation are different in these two welding processes. In K-TIG, an arc force (P_a) and the gravity (P_G) make a downward force, while the surface tension (P_s) of the melt acts upward.

The arc force is obtained from [27]:

$$P_a = \frac{\mu I^2}{8\pi}\left(1 + 2\ln\frac{R_B}{R_E}\right) \tag{2.5}$$

where R_B and R_E are the size of the arc at the workpiece and electrode, respectively, and μ is the magnetic permeability. When the summation of P_a and P_G is higher than P_s, the keyhole forms.

2.3.3 Activated-Flux TIG (A-TIG) Welding

In this method, the surface is coated with a layer of active flux before TIG welding. In case of A-TIG welding of steel, several mechanisms such as the change in the Marangoni movement cause an increase in the penetration depth [28]. This technique acts a bit differently for Al alloys. This active flux does not interfere with the electrode and increases the weld penetration without any change in its width. The active agents such as AlF_3, LiF, K_2SiF_6, and KF—AlF_3 decompose under the arc and release F^- ions, which permeate the oxide film and weaken the interface between the oxide and Al surface [29]. This facilitates the breakage and removal of the oxide layer. This F^- also reacts with the hydrogen in the arc atmosphere, which reduces the partial pressure of hydrogen in the arc atmosphere and decreases its solubility in the melt, causing the porosity to decrease [30]. When activating fluxes such as SiO_2, TiO_2, and MoO_3 are used, the arc voltage increases due to contraction of the plasma column. This changes both the microstructure of the weld and the depth-to-width ratio. The microstructure changes due to changes in G and R values [31].

2.3.4 Double-Wire MIG Welding

Double-wire GMAW, also referred to as tandem arc welding, increases the energy efficiency and reduces the pore formation. On the other hand, it causes an increase in the grain size in the HAZ and FZ due to higher heat input [32]. The energy supply of two arcs can be performed in various ways. When another wire is added to the nozzle, the temperature of the molten pool is reduced, which increases the viscosity of the melt and lessens the spatter [33]. In this state, the other wire is fed independently of the electrode wire and is connected to the negative pole to induce a force to keep it close to the wire electrode (Figure 2.8).

When both wires act as electrodes and the arc exists for both, Lorentz forces act on the droplets, whose values on the leading and trailing droplets (F_{ML} and F_{MT}) are obtained from [34]:

$$F_{ML} = j_L \times B_L = \frac{I_L}{\pi r_L^2} \times \frac{\mu_0 I_T}{2\pi D_E} = \frac{\mu_0 I_L I_T}{2D_E \pi^2 r_L^2} \tag{2.6}$$

$$F_{MT} = j_T \times B_T = \frac{I_T}{\pi r_T^2} \times \frac{\mu_0 I_L}{2\pi D_E} = \frac{\mu_0 I_L I_T}{2D_E \pi^2 r_T^2} \tag{2.7}$$

where j_L and j_T are current densities, B_L and B_T are the magnetic induction densities, I_L and I_T are the currents in leading and trailing electrodes, r_L and r_T are the radii of the electrodes, and D_E is the distance between the electrodes. These forces act

Figure 2.8 Double-wire MIG welding of Al.

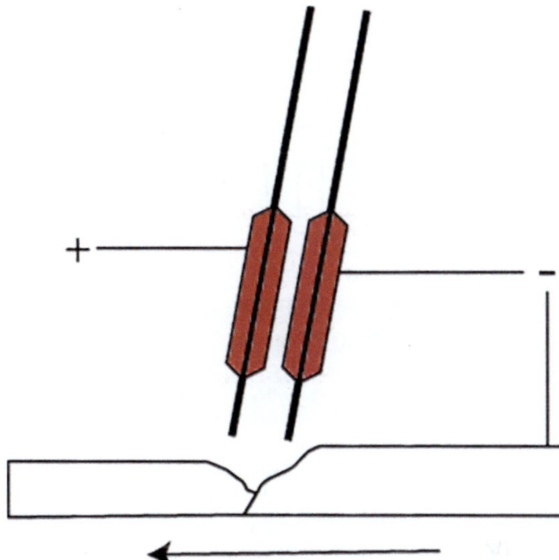

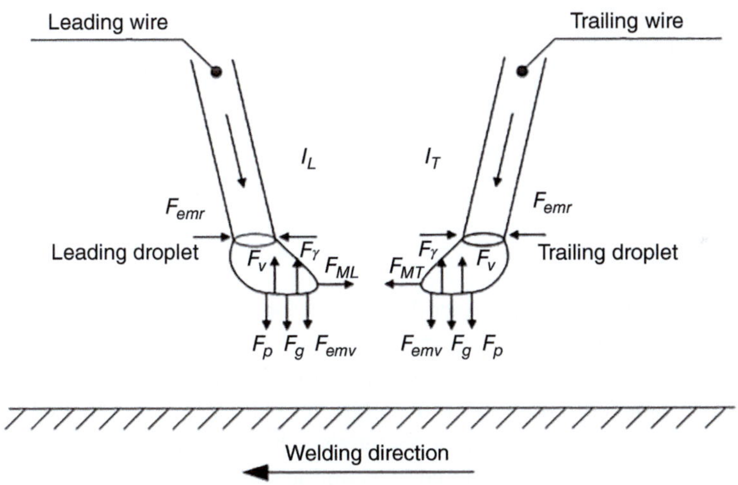

Figure 2.9 Schematic of the forces acting on the droplet during double-wire MIG welding. Source: Ref. [34]/The American Society of Mechanical Engineers.

in horizontal directions, along with the other forces that act in the vertical direction and determine detachment of the droplet from the wire (Figure 2.9). The other forces are the electromagnetic force (F_{em}), vaporization force (F_v), the surface tension force (F_s), plasma flow force (F_p), and gravity force (F_g).

When using pulsed DCs with a phase difference between the two electrodes, the force acting on the droplets is negligible as the base current is lower than the peak current, and therefore, their multiplication yields a low value, according to the above equation (Figure 2.10a). If the currents in the electrodes are in the same phase, the forces on the droplets are high and they attract each other. Due to lightweight Al,

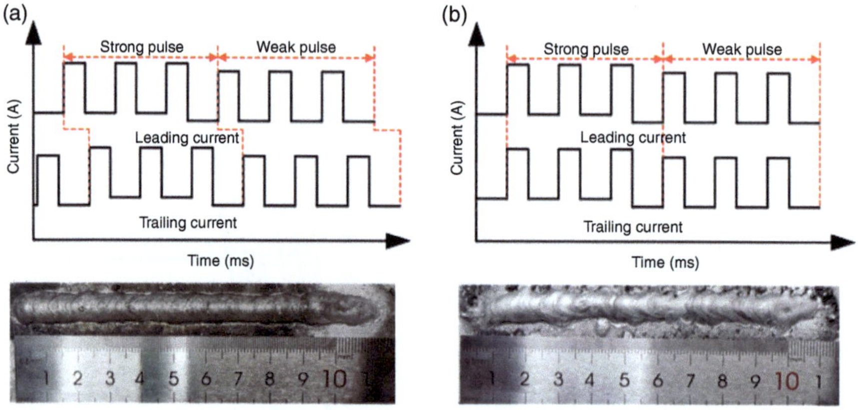

Figure 2.10 Schematic of the pulsed currents in two electrodes during double-wire MIG welding when (a) the currents have different phases and (b) the currents are in the same phase (synchronized). The below images show the corresponding weld appearance. Source: Ref. [34]/The American Society of Mechanical Engineers.

the droplets deviate from their directions and a collision occurs, which deteriorates the appearance of the weld bead (Figure 2.10b) [34].

The waveform of the current shown in Figure 2.14 is a double-pulsed GMAW, which refines the microstructure of the weld by stirring action. It also reduces the pore formation by facilitating their escape [32].

2.3.5 Indirect Electric Arc

In this technique, the fusion is not formed by the electrical arc between the wire electrode and the base metal. Instead, an extra Al plate such as AA2024 is used over the parent material where the arc is established between this plate and the electrode (Figure 2.11). The droplets produced from this plate are fed into a groove between the parent materials and melt the parent materials. This technique produces a narrower HAZ with respect to conventional MIG and is suitable for joining metal matrix composites (MMC) [35], as it partially melts the parent materials and therefore eliminates the issues raised by the presence of second phase particles in the weld pool [36]. This technique is able to join thick plates up to 12.5 mm thick by performing weld groove with a high depth-to-width ratio [37]. Also, the energy efficiency of this process is higher than conventional MIG, which is due to the hidden arc [38]. A modification to this process is used to eliminate the need for an extra plate over the parent materials, by forming a lash through machining [39]. The main issue of indirect arc welding is partial melting of the parent materials, which has been reported in all the articles related to this process. Though the root cause of this was not addressed, it is most probably due to the insufficient cleaning of the oxide layer on the abutting surfaces of the parent materials, as no arc cleaning is present.

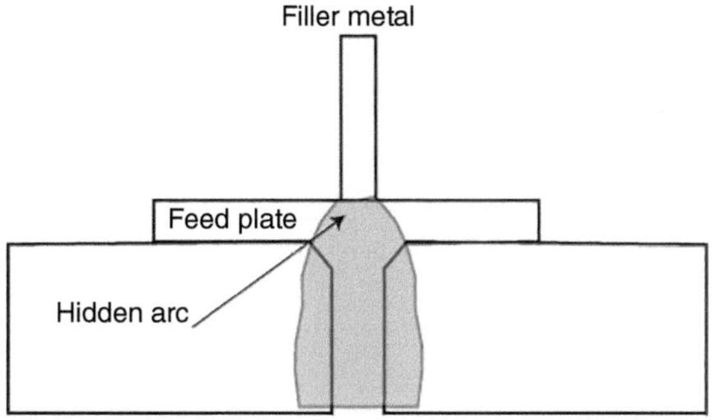

Figure 2.11 The schematic of direct and indirect electric arc MIG welding.

2.3.6 Cold Metal Transfer (CMT) Welding

This process is a modification of MIG welding in the short circuit transfer mode, with the difference that the mechanism of droplet cutting is mechanical. This cutting is performed by an innovative wire feed system and a high-speed digital control [40]. In conventional MIG welding in the short circuit mode, the droplet detachment is performed by the action of the surface tension. In the CMT process, the wire is retracted mechanically during the short circuit period. Due to its lower heat input, it is a good candidate to join Al to other metals such as steel. CMT welding can be considered as consecutively ordered spot welds that are overlapped by the torch movement [41]. It is also suitable for joining thin Al sheets, as it provides lower distortion and lower risk of burn-through. The evolution of the peak temperature in the steady state of the welding calculated on the surface of Al welded by CMT shows large oscillation of temperature between 1150 and 1800 °C, indicating why this process is suited for welding thin Al sheets [42]. The CMT process is performed in the EP mode, but it can also be done in VP. In VP-CMT welding, when the electrode is negative, the droplet size increases, which causes an increase in the deposition efficiency of the process. As the heat input decreases during Electrdoe Negative (EN), the total heat input in VP-CMT is smaller than conventional CMT where the polarity of the electrode is positive [43]. Also, the time of the droplet transfer increases when the electrode is positive due to the lower plasma drag force [44].

An example of the current waveform during VP-CMT is shown in Figure 2.12. This process is suited for gap bridging due to its higher deposition rate. The low heat input of the CMT process makes it a suitable candidate to join dissimilar materials to Al by weld-brazing technique. This is attained by controlling the IMCs thickness through the controlled heat input of the process [45].

2.3.7 Positive-Pole Plasma Welding

Like TIG welding, in reverse polarity the electrode should be cooled in plasma welding. The plasma torches using reverse polarity have a compound copper–tungsten electrode, which is water-cooled in the copper part and possesses a short tungsten electrode to lessen heating (Figure 2.13b). The dimension and shape of the electrode can be modified to use it as a micro-plasma torch intended for welding thin Al sheets (Figure 2.13c).

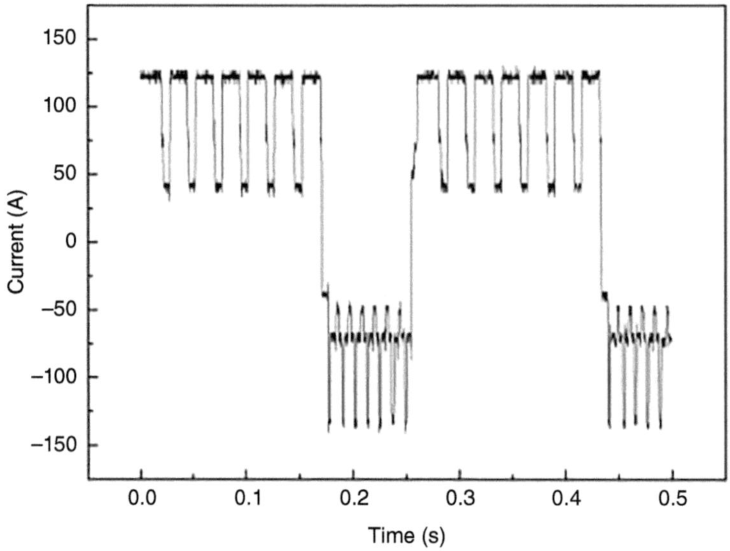

Figure 2.12 Current waveform during VP-CMT process. Source: Ref. [44]/Springer Nature/CC BY 4.0.

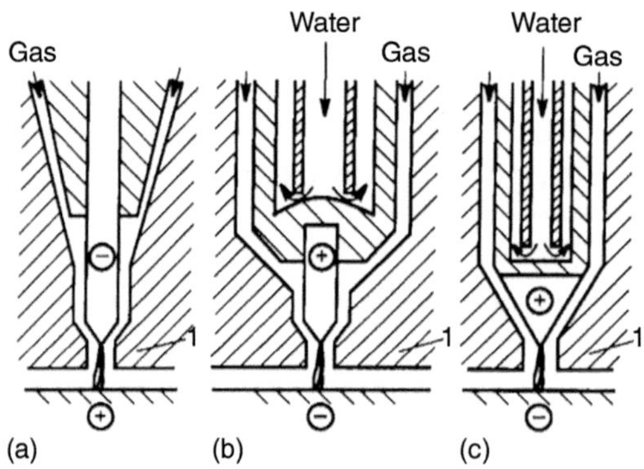

Figure 2.13 Torches for (a) straight polarity, (b) reverse polarity, and (c) micro-plasma welding. Source: Ref. [46]/Taylor & Francis.

2.3.8 Hyperbaric Welding

Hyperbaric welding refers to welding under an increased ambient pressure. MIG welding under this condition has some advantages such as reduced arc length, increased energy density, and higher depth of penetration [47, 48]. In case of Al alloys, a reduction in the pore is observed during welding under this condition, as the solubility of the gases in the liquid increases with the increase in the ambient pressure (Figure 2.14) [47].

2.4 Hybrid Welding

Hybrid welding processes are used to exploit the advantages of two processes to raise the productivity. The main complexity when using hybrid welding processes is their

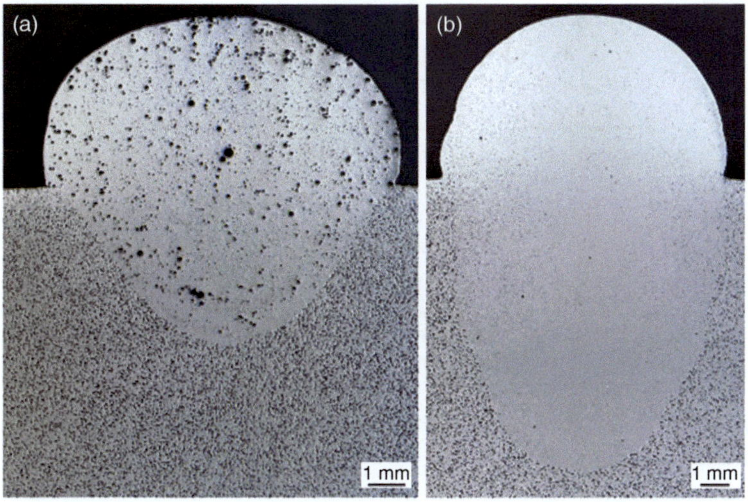

Figure 2.14 Cross sections of the weldments of Al–Mg plates welded under ambient pressures of (a) 2 bar and (b) 15 bar. Source: Ref. [47]/Springer Nature/CC BY 4.0.

interaction during welding, whether this interaction is electrical or thermal. In the following subchapters, two hybrid welding processes are explained. These processes combine two arc welding processes. Other hybrid welding processes such as laser-arc welding will be discussed in Chapter 3.

2.4.1 Plasma-MIG Welding

Combining two arc welding technologies, one with a non-consumable electrode and the other one with a consumable electrode in one melt pool, helps to use the advantages of both processes. The non-consumable electrode has a constricted arc, while the consumable one speeds up the welding. Hybrid plasma-MIG is a promising technique for welding Al with advantages such as smaller HAZ, lower distortion, and lower spatter [49]. It is considered a good alternative for hybrid laser-MIG welding, which is a costly process. In fact, plasma welding due to its comparable penetration to laser welding and its lower cost made this substitution possible. A constricted plasma arc of reverse polarity reduces the size of the arc of the consumable electrode, forming a vapor–gas channel in the weld pool without formation of a keyhole [50]. The plasma arc in front of the MIG produces a keyhole with evaporated metal, which possesses a lower ionizing potential and facilitates the MIG arc formation and increases its stability [51]. The increase in the plasma current increases the temperature and the metal vapor. The metal vapor has a lower ionizing temperature than the inert gas and therefore the MIG arc voltage reduces. An important criterion for metal transfer is achieved by which a "one droplet per one pulse" condition is established [52]. The welding pools of both processes are unified and a portion of current form MIG (EP) and plasma (EN) is connected and bypassed to the base material [52]. The MIG process in EP performs the oxide removal, and thus the plasma does not need

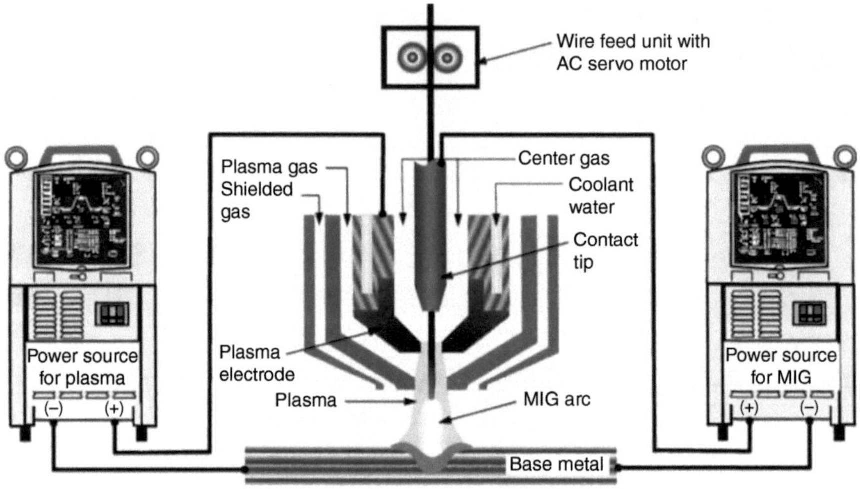

Figure 2.15 Schematic of hybrid plasma-MIG welding when the two processes are set coaxially. Source: Ref. [57]/Taylor & Francis.

to be performed in alternating or variable current modes. The two arcs can repel or attract each other depending on their polarity. Considering that the MIG process is performed in EP, the two arcs repel when the polarity of the plasma arc is EN.

Another setup of hybrid plasma-MIG, shown in Figure 2.15, is composed of an inner MIG arc coaxially installed with an outer plasma arc within a torch. The outer plasma arc is produced by an annular copper electrode. The plasma arc surrounding the MIG arc performs preheating and postheating of the wire and base metal, facilitating a higher deposition rate [53]. The temperature distribution in this process differs from conventional MIG welding because of the current shunt. By increasing the current in the outer plasma arc, the current of the MIG arc decreases, which is due to higher electrical conductivity of the plasma arc. This higher conductivity is attributed to the higher temperature and the higher amount of ionized metal vapor [53]. The highest temperature is in the outer plasma, as the current flows from the inner MIG to the outer plasma. In contrast to MIG welding, in which the highest temperature is in the center of the arc, where the temperature is more uniformly distributed, which reduces the distortion [54]. This shunt effect also decreases the temperature of the droplet, causing a higher surface tension of the droplet. This leads to an increase in the size of the droplet, which consequently reduces the frequency transfer of the droplet [54]. The main parameters of this process are the WFS of MIG, the outer nozzle diameter, and the current of plasma arc. The increase in the WFS decreases the arc length and increases the current, which consequently raises the temperature and the electromagnetic force. The size of the outer nozzle determines the current density and the electromagnetic force of the plasma arc. The electromagnetic force needs to be optimized through controlling these parameters to avoid defects such as undercut, which is common in this process [55]. The higher current in the plasma also controls the hot cracking [56].

Magnetic arc blow may occur during plasma-MIG welding of Al, which is due to asymmetric current paths in the workpiece close to the arc. This force is dependent on the current path and the poles of two arcs. This blow is governed by Lorentz force, which is obtained from [53]:

$$F_{Lorentz} = J \times B \tag{2.8}$$

where J is the current density and B is the magnetic induction. This blow also affects the flow of the melt inside the weld pool whose flow direction is dependent on the welding direction and the direction of the Lorentz force.

Plasma-MIG is less susceptible to smut formation, as the droplets released from the electrode wire have a lower temperature and therefore Mg is less evaporated [57]. The lower temperature of the droplets in plasma-MIG welding with respect to conventional MIG is due to dispersion of the MIG current to the plasma arc, which also reduces the electromagnetic pinch force, causing the droplet to be larger [58]. Also, having a high sensitivity to the plasma gas flow rate, fewer pores are formed in this process with respect to MIG [59].

2.4.2 TIG-MIG Welding

TIG-MIG welding is considered an alternative to plasma-MIG welding. This process is simpler and cheaper than plasma-MIG. The thermal efficiency of this process is lower than individual TIG and MIG processes, due to the current flow toward the TIG process (Figure 2.16b). Also, the current in the TIG process should be higher than MIG to stabilize the process. The enhanced vapor metal decreases the ionizing potential, which manifests itself in the characteristic curves (Figure 2.16c) [60].

Narrow-gap welding (NG welding) is a technique used to weld heavy plates using a low heat input, which results in a narrow HAZ and small deformation. The main challenge of this technique is incomplete penetration in the side walls [61]. TIG welding in front of the of MIG stabilizes the arc of the MIG by preheating the droplets

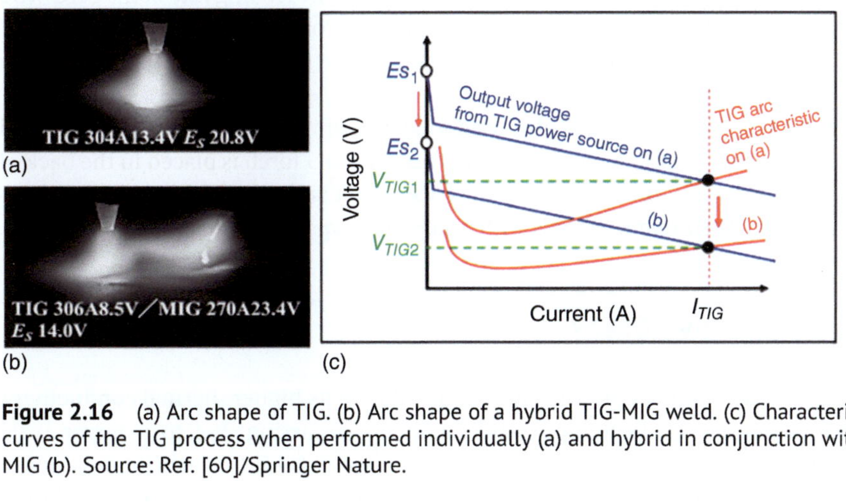

Figure 2.16 (a) Arc shape of TIG. (b) Arc shape of a hybrid TIG-MIG weld. (c) Characteristic curves of the TIG process when performed individually (a) and hybrid in conjunction with MIG (b). Source: Ref. [60]/Springer Nature.

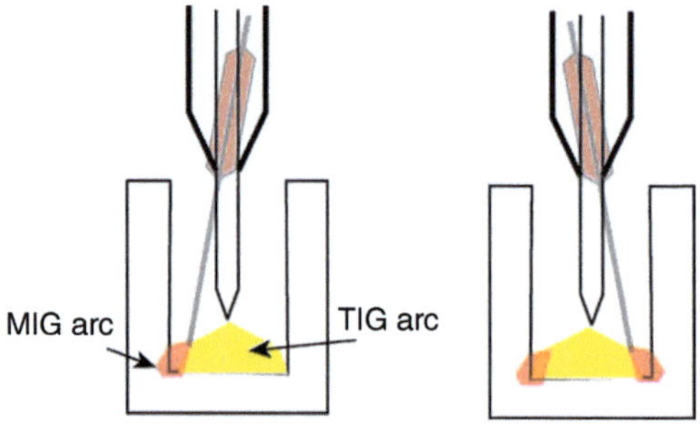

Figure 2.17 Swing TIG-MIG welding for narrow-gap welding.

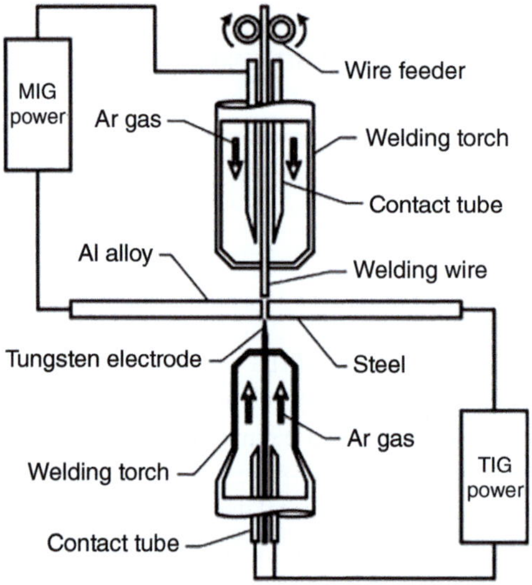

Figure 2.18 Schematic of hybrid TIG-MIG weld-brazing used to join Al to steel. Source: Ref. [62]/Springer Nature.

and workpiece. Swinging the TIG torch is an effective method to ensure the complete penetration in the sidewalls [61]. Swinging the TIG torch inside the gap is performed with an oscillator toward the sidewalls in a way to avoid the direct contact of the tungsten electrode with the sidewalls (Figure 2.17). The feeding rate of the MIG wire needs to be adopted with the TIG current to make a proper distance between the MIG wire and the bottom of the narrow gap, ensuring the arc of the MIG does not blow to the sidewalls [61].

Hybrid TIG-MIG welding is also used to join dissimilar materials such as Al and steel through weld-brazing technique. In this way, TIG torch is placed in the back of the weld to provide sufficient heat in the back side (Figure 2.18). This facilitates joining of thicker dissimilar materials with a narrower gap in the butt joint configuration.

2.5 Protective Gases

He gas is added to Ar by 50% in TIG welding due to its higher thermal conductivity, which increases the heat input and facilitates degassing due to a larger weld pool [13]. The type of the gas can also influence the HAZ through the heat input delivered to the workpiece. In this regard, Ar leads to a higher hardness and finer microstructure in the HAZ [63]. Higher amounts of He make the arc unstable. The addition of a small amount of nitrogen in MIG welding or oxygen in TIG welding makes the arc very stable, letting higher speed of welding and consequently lower distortion [8]. They also help to reduce the movement of the arc and improve its concentration precision when the electrode is positive [13]. They also help to increase the penetration depth [32].

2.6 Filler Material

During arc welding of Al alloys, filler materials are used with several considerations. Mechanical properties, corrosion resistance, and weldability are the main considerations for filler materials selection. The main issue in weldability of Al alloys is hot cracking, which, as discussed in Chapter 1, is primarily dependent on the chemical composition. While the chemical composition in the PMZ is not changeable, it can be changed in the FZ through the selection of the filler material whose contribution is dependent on the dilution. The other issue for filler material selection is to avoid brittleness of the weld through the excessive formation of Mg_2Si compounds. This occurs when the base materials with Mg content higher than 2 wt% will be welded by filler materials with a high Si content such as $AlSi_{10}$. Al—Mg alloys when possessing high Mg contents (>3%Mg) are prone to inter-granular corrosionwhen stored at temperatures >65 °C. This is due to the formation of a network of Al_3Mg_2 precipitates along the grain boundaries [64]. Therefore, the AlMg alloys (5xxx) need to be welded with filler materials that keep their Mg content below 3%. This is often obtained by using a filler material similar to the base material. When the filler material and the base material are the same, the process is referred to as homogeneous welding, and when they are different, the process is heterogenous welding. When no filler material is used, for example by TIG welding, the process is referred to as autogenous welding. Al alloys of 6xxx and 7xxx cannot be welded autogenously as their chemical composition is in the susceptible range for hot cracking [65, 66]. For 7xxx, one proper filler material is AA5xxx, though AA4xxx also yields a weld without SC [67]. A combination of several factors such as the filler materials, the process parameters, and the joint geometry affects the chemical composition of the joint with regard to the dilution and the chemical composition of the FZ. The viscosity of the melt is determined with the temperature and chemical composition. A higher viscosity makes the backfilling process (refer to Chapter 1 for permeability during SC) difficult and raises the solidification cracking susceptibility (SCS). Heavy elements such as Cu in the melt raise the viscosity and consequently the SCS [68]. Table 2.1 summarizes the filler materials used to join various Al alloys by arc welding processes.

2.7 Grain Refinement in the Weld

Grain refinement in the FZ improves the mechanical properties such as the yield strength and toughness. It also raises the crack resistance of the weld [74]. One effective method to refine the microstructure of the weld is increasing the

Table 2.1 Filler materials used to join various Al alloys by arc welding processes.

Parent metal 1	Parent metal 2	Process	Filler material	Protective gas	Reference
AA5754	AA6013	GMAW	ER5356	Ar	[69]
AA5083	AA6082	GTAW	ER4043		[70]
AA6061	AA6061	GTAW	ER4043	50%Ar + 50%He	[71]
AA2124	AA7075	GTAW	ER5356		[72]
AA2024	AA6061	GTAW	ER5183	Ar	[68]
AA6063	AA6063	GTAW	ER5356	Ar	[73]
AA5059	AA5059	GTAW	ER5556	Ar	[9]
AA2219	AA2219	GTAW	ER2325	Ar	[5]

nucleation sites. Solidification nuclei are particles of micrometer size that are not soluble in the molten Al. The effectiveness of nuclei particles during cooling of the melt depends on their chemical composition, size, form, and crystal structure. The most efficient nuclei in Al melt contain Ti, B, or both [75]. Examples are TiB_2 and Al_3Ti. Practically, the rod-shaped grain refiners of $AlTi_5B$ are added into the molten Al during casting to achieve a fine grain structure. The mechanism of grain refining is based on a peritectic reaction between the molten Al and grain-refining particles, which takes place above the liquidus temperature [76]:

$$\mathrm{Al\,(L)} + \mathrm{Al_3Ti} \rightarrow \mathrm{Al\,(S)} \tag{2.9}$$

The effectiveness of these particles to produce equiaxed grains depends on the number of nucleation sites per volume (N_0). Also, an undercooling (ΔT_N) is necessary for these particles to be activated. Ideally, this undercooling is lower than the available undercooling in the solidification front (ΔT_C). When the temperature gradient (G) is low, more time is available for the process of nucleation and growth of the equiaxed grains, before they meet the columnar solidification front. Therefore, for a given set of conditions, there is a critical temperature gradient over which no grain refining occurs. The following relationship yields the critical temperature gradient for transition of columnar to equiaxed grain structure [76]:

$$G < 0.617\, N_0^{1/3} \left(1 - \frac{\Delta T_N{}^3}{\Delta T_C{}^3}\right) \Delta T_C \tag{2.10}$$

When the above equation is met, about 50% of the grains will become equiaxed. According to this relationship, when the temperature gradient is high, higher nucleation particles are needed to promote the equiaxed formation as well as the grain refining. Also, this equation can be used to realize the effect of the welding process and parameters on the formation of fine and equiaxed grains when using nucleation particles. As an instance, during laser welding of Al, a higher content of nucleation particles is needed to attain a fine and equiaxed structure in the weld than TIG welding. Considering the higher solidification rate in the weld pool in laser welding than

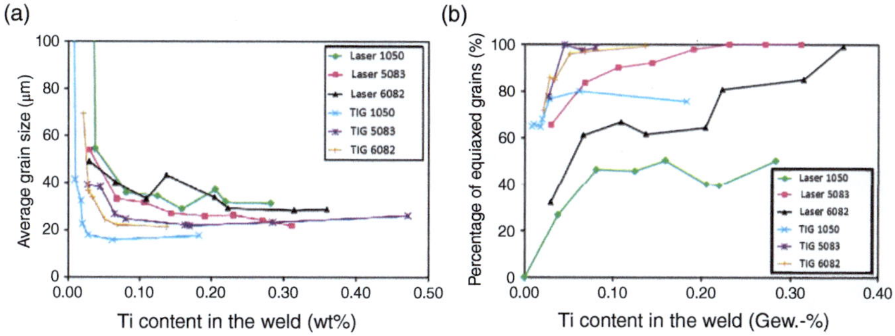

Figure 2.19 (a) Effect of Ti content on the (a) average of grain size and (b) the area fraction of equiaxed grains in the weld zones of three Al alloys welded by TIG and laser welding. Source: Ref. [76]/DVS Verlag GmbH.

TIG welding, a contrary result is expected. Here, the temperature gradient plays a more important role than the solidification rate. In TIG welding, a larger weld pool exists than in laser welding, which results in a lower G. According to the above equation, at the same content of nucleation particles, a finer and more equiaxed grain structure develops during TIG welding than in laser welding [76].

Other than the number of nucleation sites, the effectiveness of the Ti on grain refinement depends on the alloying elements of Al. By increasing the alloy elements content, the available nucleation particles decrease due to the chemical reaction between TiB_2 and Al_3Ti from one side and the alloying elements from the other side [77]. That is why in Figure 2.19a the average grain size during TIG welding of AA1050 is lower than the other Al alloys. Also, a higher grain size is seen during laser welding than TIG in this figure. In Figure 2.19b is shown the same graph for the area fraction of the equiaxed grains. Two points are derived from this figure: First, the area fraction of equiaxed grains is higher in TIG welding than in laser welding. Second, the area fraction of equiaxed grains in AA1050 is lower than the other Al alloys, showing the importance of alloying elements in the development of equiaxed grain during solidification.

The previously explained mechanism for why Ti addition refines the grain structure during solidification is based on the paradigm of nucleation sites on the surface of the nucleants. Another mechanism is based on the paradigm of solute undercooling-driven mechanism [78]. In contrast to the nucleant theory, Ti and other elements refine the microstructure not only in the solid state, but also when they are soluble in the liquid. The alloying elements act in two ways: First, they dissolve in the melt and then they segregate in front of the solidification interface, which restricts the growth rate of the cells and provides more time for nucleation event to occur. Second, and more important, this segregation provides further constitutional undercooling, which activates the nucleants in the solidification front (Figure 2.20) [79]. In this way, a growth restriction factor (GRF) is defined, which measures the contribution of each alloying element in growth restriction [78]:

$$GRF = mc_0(k - 1) \tag{2.11}$$

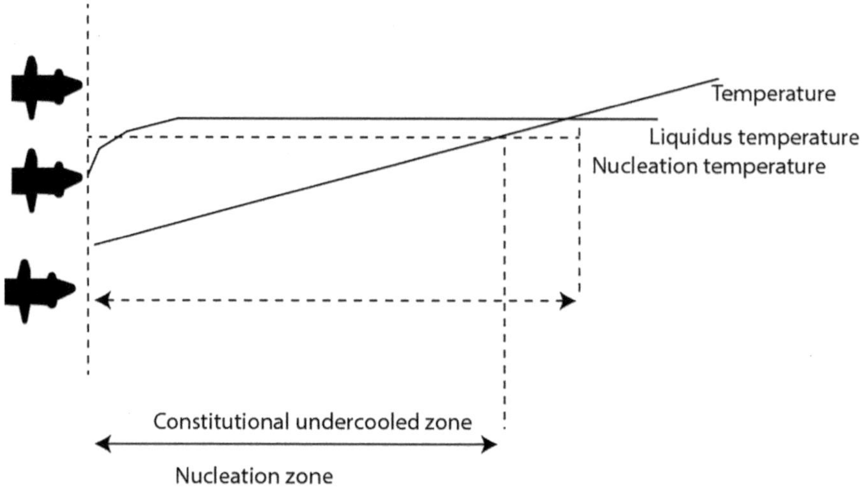

Figure 2.20 Schematic of the constitutional undercooling resulting from the segregation of the solute alloying elements in front of the solidification interface where the nucleants are activated (edited). Source: Ref. [78]/Springer Nature.

Table 2.2 Values of GRF for three Al alloys.

Al alloy	GRF value
AA1050	3.2
AA5083	10.6
AA6082	7.4

Source: Adapted from [76].

where m is the gradient of the liquidus line in the phase diagram, c_0 is the concentration of the solute, and k is the partition coefficient between the equilibrium concentrations of the solid and liquid at the growing interface $\left(k = \frac{C_s}{C_l}\right)$ at the interface temperature.

The quantity of GRF is summed over all the alloying elements in the liquid. In this way, the effect of the alloying elements on the area fraction of equiaxed grains can be explained. Table 2.2 presents the values of GRF for three Al alloys without considering the master alloys of Ti and other nucleants, as it is very difficult to measure how much Ti is dissolved in the melt. GRF is lowest for AA1050, which explains well why the area fraction of the equiaxed grains in AA1050 is the lowest. The type of the filler material also affects the grain size in the FZ [67]. AA5xxx filler material to weld AA7075 by TIG welding yields a finer structure in the FZ with respect to AA4xxx filler materials.

The morphology and size of the grains in the FZ can be engineered using various techniques. Understanding the physics of grain engineering by these methods helps to use them during arc welding processes to obtain the desired microstructure.

Ultrasonic-assisted arc welding processes use a physical phenomenon made by ultrasonic cavitation through which an instantaneous high pressure (P_{max}) is induced according to the following equation [80]:

$$P_{max} = P_\gamma \left[\frac{P_m(\gamma - 1)}{P_\gamma} \right]^{\frac{\gamma}{\gamma-1}} \tag{2.12}$$

where P_γ is the vapor pressure inside the cavitation bubble, P_m is the acoustic pressure amplitude on the surface of cavitation bubble, and γ is the surface tension. The large pressure produced by the collapse of the bubbles increases the freezing temperature (T_m) of the alloy according to Clausius–Clapeyron equation:

$$\frac{dT_m}{dp} = \frac{T_m(V_L - V_s)}{L} \tag{2.13}$$

where L is the latent heat of solidification, and V_L and V_S are the specific volumes of the liquid and solid, respectively. The right term is positive and hence the freezing temperature increases with increasing pressure. This causes the undercooling degree to increase and hence the heterogenous nucleation accelerates [80]. The mechanical impact of the ultrasonic vibration has no effect on the grain refinement, and it is the ultrasonic energy that causes this [81]. Figure 2.21 shows the microstructure of a pure Al welded with periodic application of ultrasonic vibration. The equiaxed region with fine grains was obvious when welding was performed with ultrasonic vibration (between the dotted yellow lines).

Stirring by external magnetic field is another strategy used to refine the grains during arc welding processes [83]. This occurs through the changes in the flow of the melt and the cooling rate in the FZ. It also helps to repress the SC in the FZ [84].

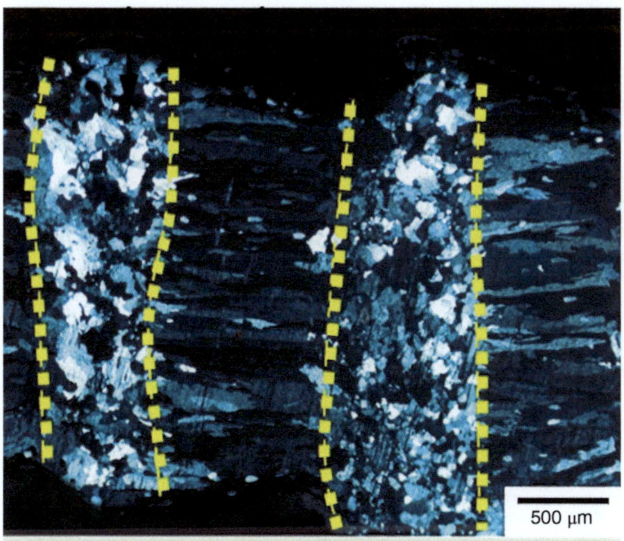

Figure 2.21 Microstructure of a pure Al TIG-welded under periodic application of ultrasonic vibration. Source: Ref. [82]/with permission of Elsevier.

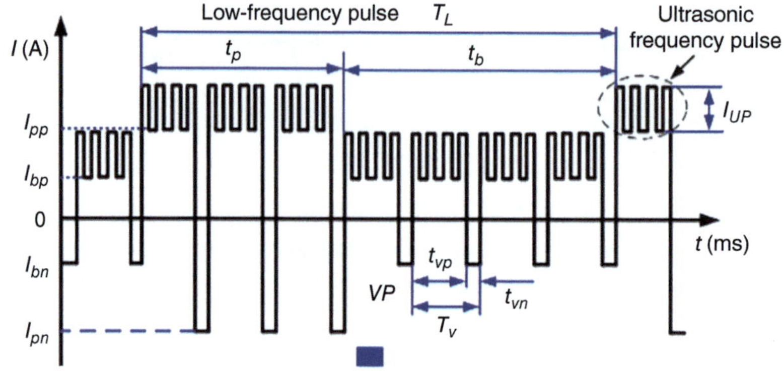

Figure 2.22 Wave current of DP VP-TIG. Source: Ref. [87]/Elsevier/CC BY-NC-ND 4.0.

Current pulsing also refines the grains in the FZ [85]. The ultra-high-frequency current generates electromagnetic force that stirs the FZ during welding. This provides sites for heterogenous nucleation, which promotes fine equiaxed grain formation [86]. The pulse frequency needs to be optimized to obtain desirable grain structure. A modulated double-pass VP-TIG welding (Figure 2.22) consisting of a low frequency (0.5–10 Hz) and an ultrasound frequency (20–80 kHz) further improved the refining process and the segregation of eutectic compounds [87]. During the low-frequency pulse, when the current is at peak (t_p), a keyhole forms (Figure 2.23a), and when the current is at its base (t_b), the keyhole disappears letting the melt to fill the hole (Figure 2.23b). This procedure improves the fluidity of the melt. The ultrasonic waves generated during the corresponding pulse cause cavitation, which improves the fluidity of the melt and facilitates breakage of the dendrites in the rear of the weld pool. These dendrites act as nucleation sites and refine the grains.

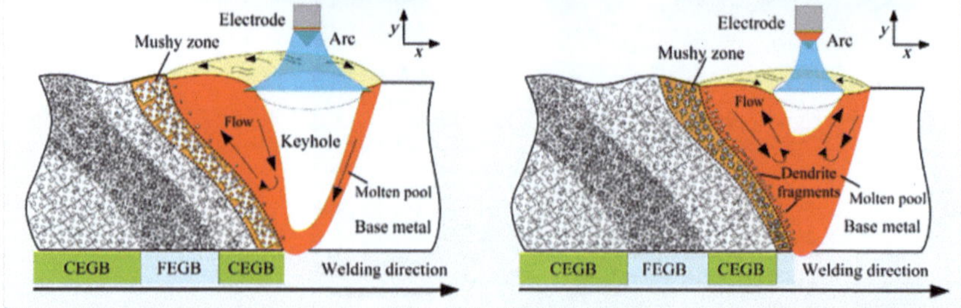

Figure 2.23 Evolution of the microstructure during double-pass VP-TIG welding of Al during (a) peak stage of current and (b) base stage of the current in the low-frequency zone. Source: Ref. [87]/with permission of Elsevier.

Arc oscillation is another method that refines the microstructure of the weld. Amplitude, frequency, and the oscillation direction are the most important parameters adjusted to refine the grains. The main mechanism of grain refinement is dendrites fragmentation [74]. The effectiveness of this method depends on the chemical composition of the weld, as some elements such as Si in 6xxx alloys increase the viscosity of the melt and make the flow more difficult [74].

2.8 Grain Structure

During fusion welding, a dendritic structure exists at the fusion boundary due to thermal conditions discussed in Chapter 1. This dendritic structure grows epitaxially from the grains of the PMZ region and becomes coarse when its growth rate is more than their formation rate in the fusion line. Figure 2.24 shows the grain structure of the weld of an AA6061 welded by TIG welding. The coarse dendrites are obvious near the fusion boundary.

For Al—Li alloys, the situation is different, as a fine equiaxed zone (FQZ) exists in the fusion boundary of the FZ (Figure 2.25b). This region deteriorates the performance of the weld, as it is the softest region in the weld. This softening is due to segregation of elements between the grains in the form of eutectic and IMC phases, which depletes the grains from the alloying elements and reduces the precipitation inside the grains (Figure 2.25c) [88]. This region is also weak in fatigue loading and breaks inter-granular due to the presence of weak phases between the grains [89]. The presence of Zr refines the grains and coarsens the eutectic phase, making this region even more weakened. The presence of Li and Zr inhibits the epitaxial growth from the fusion line. This is a region of the FZ close to the fusion boundary where the melt is stagnant and is named the unmixed zone. The mechanism of equiaxed grain formation is heterogenous nucleation. The favorable sites for heterogenous nucleation are Al_3Zr precipitates in the base material which act effectively in the temperature range of 630–640 °C as they neither dissolve nor detach from the fusion line. Li reduces the surface tension of molten Al (σ_{LS}) and therefore reduces the energy barrier for heterogenous nucleation (ΔG) according to the heterogenous nucleation theory [89]:

$$\Delta G = \left(\frac{16\pi{\sigma^3}_{LS}}{3(\Delta G_V)^2}\right)\left(\frac{2 - 3\cos\theta + \cos^3\theta}{4}\right) \tag{2.14}$$

where ΔG_V is the Gibbs free energy difference between liquid and solid per unit volume and θ is the contact angle. To avoid the formation of this region, one strategy

Figure 2.24 (a) Macrograph of the weld of a 6061 alloy. (b) Grain structure within the weld zone. Source: Ref. [74]/Sage Publications.

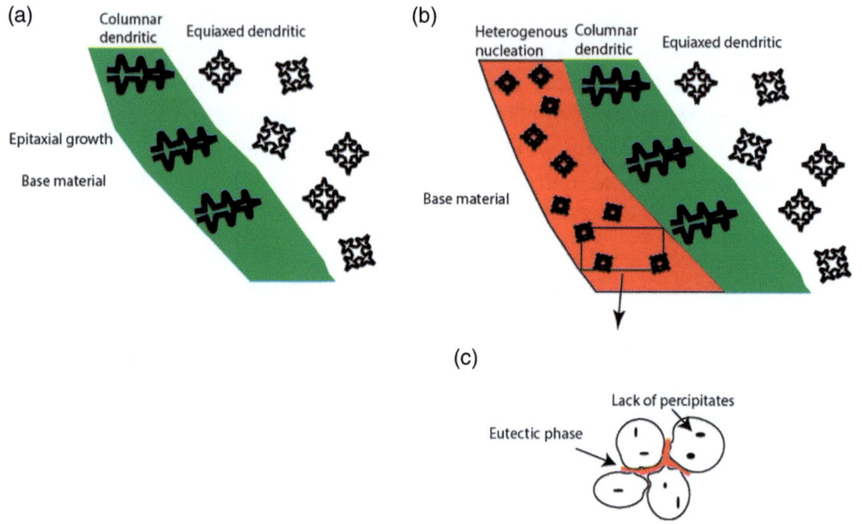

Figure 2.25 Schematic diagram of (a) conventional epitaxial growth and (b) fine equiaxed structure formation at the fusion boundary. (c) Depletion of grains interiors from alloying elements and consequent lack of precipitates after solidification.

is to eliminate the nucleation sites by rapid mixing in the weld pool [21]. Another condition for FQZ is that the base material should be homogenized.

When doing multi-pass welding, the subsequent passes have thermal effects on the preceding passes. For instance, in a weld with three passes, the third pass causes solutioning of the second pass. The first phase, which was previously under solution effect of the second phase, goes under aging treatment by the heat of the third pass [90]. This causes a nonuniform hardness distributionin the weld zone. The penetration of each pass, which is directly connected with its heat input, determines the extent of its influence over the previous passes and the weld properties [91].

2.9 Defects During Arc Welding

In this section, only the defects that are originally metallurgical are discussed. Other defects such as lack of fusion or penetration have non-metallurgical bases.

2.9.1 Segregation

Segregation of alloying elements in the weld zone can give rise to SC under some circumstances, as discussed in Chapter 1. Even if the SC does not occur, the segregation deteriorates the mechanical properties of the joints, causing an inconsistency in the mechanical properties of the joint [92]. This segregation is more obvious in the fusion boundary where an elongated dendritic structure exists. The segregation may appear as an eutectic structure along the boundaries of these grains, acting as the initiation site for fracture during loading [92]. The existence

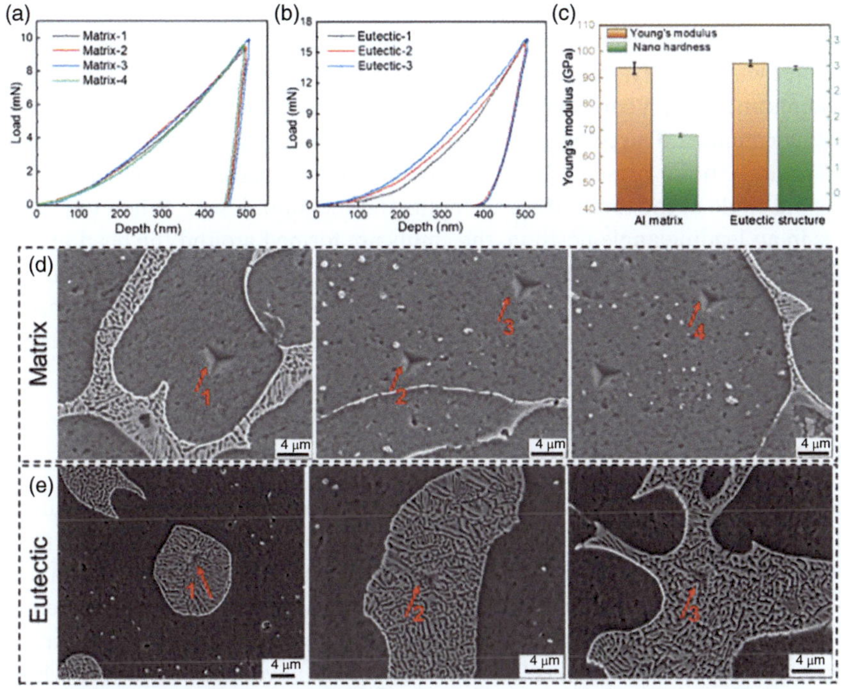

Figure 2.26 Nanoindentation test of the (a) matrix and (b) eutectic structure in PMZ of an AA2024 alloy welded by TIG. (c) Extracted values of hardness and Young's modulus. Indentation in the (d) matrix and (e) eutectic structures. Source: Ref. [93]/with permission of Elsevier.

of eutectic structure in the PMZ of some Al alloys such as AA2024 causes a huge reduction in the strength of the joint. The fracture strain of the eutectic structure is much lower than the matrix (0.03 for eutectic and 0.55 for matrix) [93]. In fact, the eutectic structure is brittle and the crack initiates perpendicular to the loading direction. The strain-localized bands form between the eutectic structures due to incompatibility in deformation [93]. Figure 2.26a and b show the nanoindentation test results made on the matrix and the eutectic structure in PMZ of an AA2024 alloy welded by TIG, respectively. Figure 2.26c shows how harder is the eutectic with respect to the matrix. The indentations on the matrix and eutectic structure are shown in Figure 2.26d and e, respectively.

2.9.2 Smut

Smut spots are colored deposits on the surface of the weld made by MIG, which contain the elements with high-pressure vapor such as Mg, Zn, and Li [55]. The mechanism of formation of smut is condensation and oxidation of the metallic vapors around the bead where the temperature is lower than the arc atmosphere. The metallic vapors originate from the arc atmosphere where they escape outside the shielding gas and are transformed by the consumable electrode to the bead surrounding [94]. The chemical composition of the smut consists of Al, Mg, oxygen, and carbon, the latter originating from the graphite lubricant used on the surface of the filler wire [53]. Though the smuts do not impair the mechanical properties of the joints and are not considered as defects, their formation should be controlled for aesthetic matters. The color of the smuts is black when Mg is present and is gray when Mg and Zn are present. The presence of smuts depends on several factors such as the angle of the torch, the arc length, and the joint design. When the MIG torch is in an leading angle position, the smuts are formed around the bead in the toe of the weld, and when the torch angle is trailing, the smuts cover the entire surface of the weld. When the arc length is short, the number of smuts decreases. The smuts are lower in number when welding groove welds and fillet welds than welding the square butt joints and bead on plate [94].

2.9.3 Pores

The pore formation during arc welding of Al originates from hydrogen, which causes metallurgical pores. Other mechanism of pore formation pertaining to other processes are described in Chapter 3. The metallurgical pore formation depends on the process parameters. In case of TIG welding, the current and the torch speed determine the pore formation. These two factors determine the total heat input during the process. The pore formation is less likely to occur at lower heat input [95], while it is expected that at low heat input the escape of the bubbles is less probable. This is because the high welding speed at low heat input necessitates a high current, and this influences the dilution and melt flow, which influence the pore formation [96]. Besides the heat input, vibration of the molten pool with ultrasonic can reduce the pores during welding Al [97].

2.9.4 Hot Cracking

The main challenge in using arc welding processes for Al alloys is hot cracking in 7xxx and 2xxx alloys. Hence, in the aerospace industry, riveting and bolting are used to join these materials. The FSW process has been developed to weld these materials without hot cracking, but its application for welding these materials is limited due to difficulties in welding of complex geometries. Arc welding processes are easy to implement for welding during both fabrication and assembly, and hence the interest still exists to solve the issues related to hot cracking. One strategy to solve this problem is modification of the chemical composition of the weld zone by using a proper filler material, so the composition of the weld falls outside the susceptible range. The main drawback of this strategy is the change in other properties of the weld with respect to the base material, which deteriorates the integrity of the structure. The change in the composition of the weld changes the mechanical properties and even the corrosion resistance. It also changes the response behavior of the weld zone to the heat treatment after welding, causing the base material and the weld zone to show different properties after heat treatment. The strategies to use the filler material similar to the base material need special considerations to avoid the hot

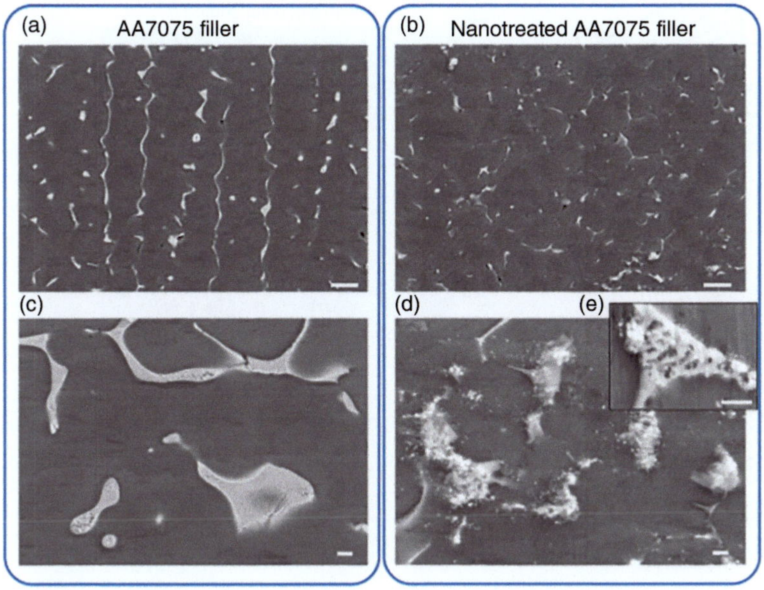

Figure 2.27 SEM images of the weld zone of an AA7075 welded with (a and c) 7075 filler material and (b and d) TiC-containing 7075 filler material and (e) Higher magnification image of broken eutectic. Source: Ref. [99]/Springer Nature/CC BY 4.0.

cracking in both the welding zone and the PMZ in the forms of SC and liquation cracking, respectively. As mentioned in Chapter 1, the higher solid fraction of the FZ with respect to PMZ adjacent to the fusion boundary in the final stages of its solidification induces tension in PMZ and causes the liquation cracking [98]. The usage of similar filler material intrinsically alleviates this problem, and no liquation cracking occurs in the PMZ [99].

Filler pastes containing nanoparticles of TiC applied onto the joint seam before welding eliminated the SC in TIG welding of 7075 alloy [66]. This is achieved by changing the morphology of the grains from columnar to equiaxed. In another approach, TiC nanoparticles were incorporated into the filler wires of AA7075 to weld Al sheets of AA7075 by TIG welding [99]. The elimination of SC by using TiC nanoparticles is attributed to the change of segregation pattern of the second phase at the grain boundaries (Figure 2.27a and b) as well as the accumulation of TiC nanoparticles in these regions which alter the eutectic structure from a lamellar one to a broken one (Figure 2.27c and d). This mechanism along with the refined equiaxed grains further reduces the SC susceptibility explained before. This method has several advantages for welding non-weldable Al alloys, such as elimination of SC, astonishing mechanical properties due to the fine microstructure and dispersion of TiC nanoparticles, and similar response to the PWHT.

The addition of TiC into the filler material extends the choices of various Al alloys for welding. The formation of refined equiaxed grains in the weld zone of various Al alloys is reported, such as AA2024, AA5083, AA061, and AA7075, where no SC was observed [100].

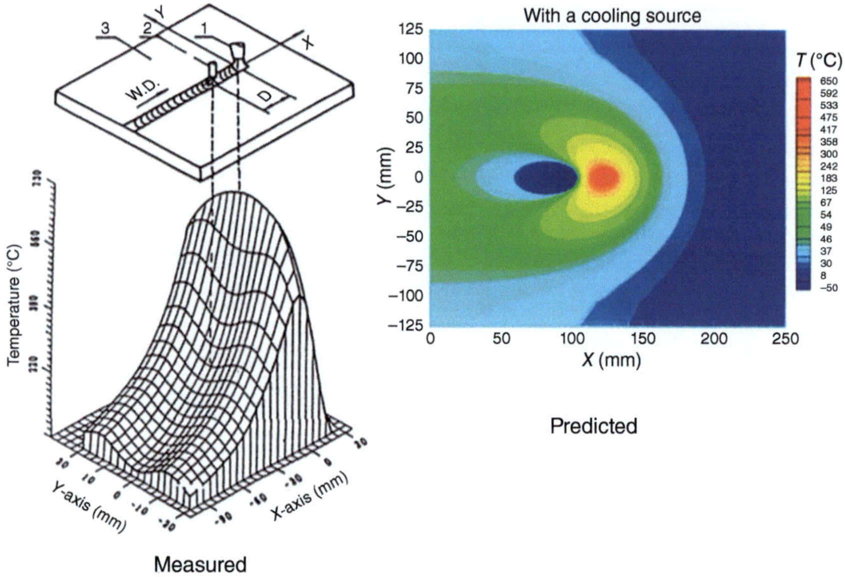

Figure 2.28 Measured and calculated temperature distribution using a trailing heat sink. Source: Ref. [103]/American Welding Society.

Some Al alloys such as AA2219 are prone to liquation cracking or loss of strength due to segregation in the PMZ. The swing technique performed on the final cap weld can alleviate this problem by reducing the stress concentration in the toe of the weld [101]. Using pulsed current welding is another way to overcome hot cracking during fusion welding. One reason is the lower heat input during pulsed welding, which refines the grains and hence lowers the susceptibility to hot cracking [102]. Trailing heat sink mitigated the hot cracking during TIG welding of AA2024 by reducing the mechanical strain within the BTR [103, 104] (Figure 2.28).

2.10 Welding of Dissimilar Metals

The direct mixing of the melts of dissimilar metals, for example Al and steel, produces massive IMCs that are brittle and degrade the joint strength. Hence, arc welding processes are usually used to perform weld-brazing instead of welding by which only the lower-melting-temperature metal (Al) is fused and spread over the solid surface of the other metal (steel) to make a metallurgical bonding. This is different from brazing in which the filler material spreads over both surfaces in the solid state by the capillary effect, filling a very narrow gap between the two sheets. To improve the wettability of the melt over the surface of steel, the surface of steel is galvanized to form a zinc layer. Also, hot-dip aluminizing can be used to form an Al layer over the steel to improve wettability. Though the wettability of Al over zinc is better, Al layer produces lower IMCs during weld-brazing [105]. Weld-brazing is performed in a lap configuration to facilitate the spread of the molten material over the surface of

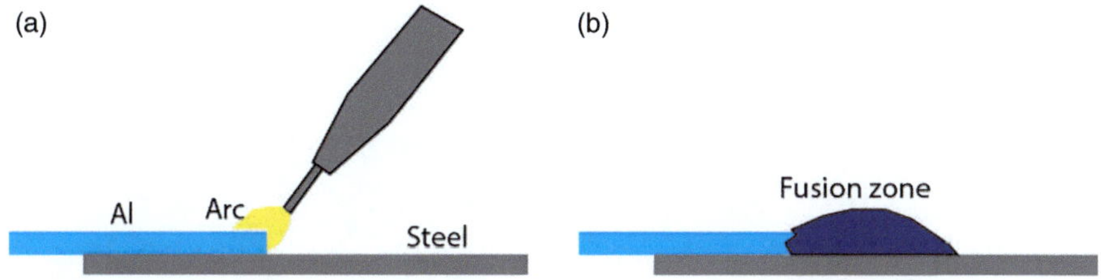

Figure 2.29 (a) Schematic of weld-brazing of Al to steel where the low-melting material is placed on top. The torch is directed toward Al. (b) Schematic of the joint produced by weld-brazing.

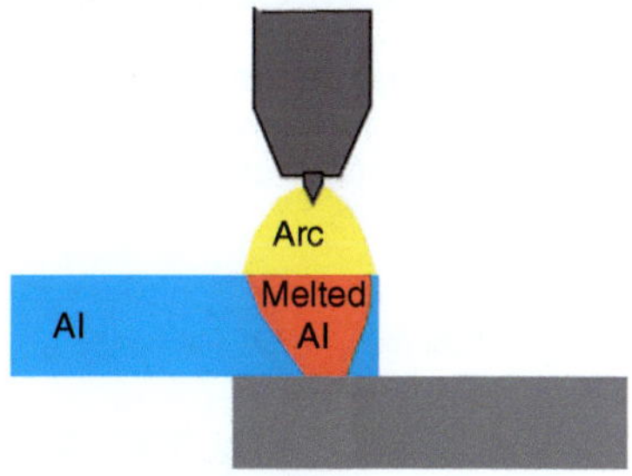

Figure 2.30 Schematic of weld-brazing of Al/Cu with Al placed on top using plasma arc welding.

the other material. A precise selection of the filler, heat input, and the torch angle is critical to obtain a good joint. The arc is directed mainly toward Al to avoid excessive heating of the steel and quick evaporation of the zinc layer (Figure 2.29) [106].

In this approach, the low-melting material is placed on top to facilitate its melting and its spread over the other material which is in the solid state. Moreover, a low portion of the arc heat is directed to the lower material. In another approach, an intense heat source such as plasma arc can be used to heat only Al, causing it to melt down to the interface and transfer the heat to the bottom layer (Figure 2.30) [107].

Reverse approach is also used to do weld-brazing of Al to other materials such as Ti, wherein the low-melting material is placed at the bottom (Figure 2.31a and b) [108]. In this state, the arc is targeted into Ti to melt Ti, but the extent of its melting does not reach Al. In fact, the heat is transferred through Ti to Al to melt the Al, which possesses a lower melting point (the melting point of Ti alloy is 1668 °C and for Al alloys is 660 °C), without Ti to melt at the interface. Figure 2.31c shows a high-magnification image of the interface which shows how far is the HAZ of the Ti alloy to the interface, leaving an unfused zone of Ti close to the interface.

CMT welding can also be used to do weld-brazing. While the low heat input of CMT is an advantage to control the IMCs thickness, it has limitation for weld-brazing of thick Al sheets to steel, as the heat dissipation in the thick Al sheet causes insufficient melting of Al and its wetting on the surface of steel. A new approach that combines pulsed-MIG and CMT was used to join thick Al sheets to steel. The pulsed MIG compensates for the heat dissipation in thick Al sheets [109].

Weld-brazing for butt joints with a closed square groove is not feasible by using conventional arc welding processes, as the molten material cannot spread over the

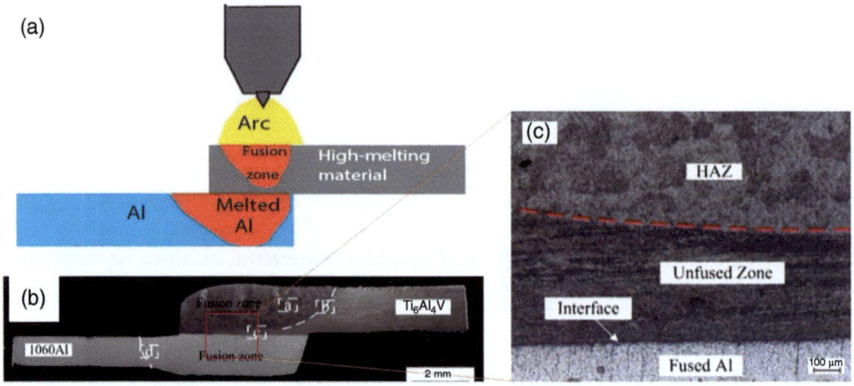

Figure 2.31 (a) Schematic of plasma arc weld-brazing of Al to Ti alloy where the low-melting material is placed at bottom. (b) Cross section of the joint. (c) High-magnification image of the interface of Al/Ti. Source: Ref. [108]/with permission of Elsevier.

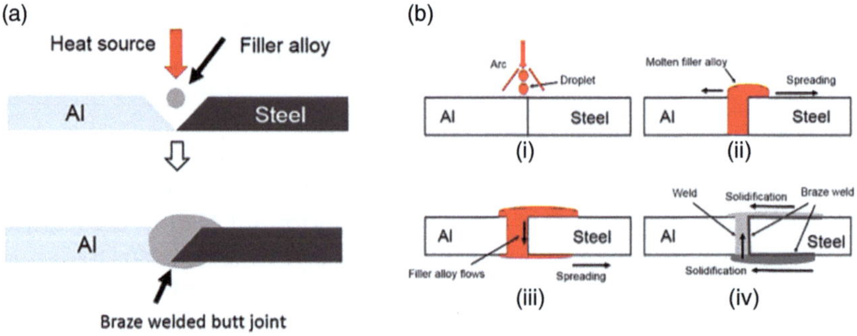

Figure 2.32 (a) Edge preparation for weld-brazing in butt configuration. (b) Weld-brazing of Al/steel by CMT. Source: Ref. [41]/Springer Nature.

entire surface of the other material, especially the root and backside. For this to happen, a very high heat input is required, which causes excessive growth of the IMCs. Hence, weld-brazing in butt joint usually needs a bevel to provide a surface where the melt of the low-melting material can spread over the other material (Figure 2.32a). CMT can be used to perform weld-brazing of thin sheets in close square groove by pointing the arc into Al (Figure 2.32b). By CMT, a high start current melts Al and makes a full penetration into Al without melting steel. The Al melt spreads over the surface of steel and makes a metallurgical joining.

In another method to alleviate the problem of insufficient heat in the backside, weld-brazing in butt configuration was performed by a combined TIG-MIG welding procedure to weld Al to Ti [110]. The heat source in the back of the joint facilitates the spreading of the molten material in this region. The resultant joint shows no sign of Ti melting. The filler wire used in the MIG process was ER4043, which filled the gap between the sheets. The only determinant factor in the joint performance under loading is the IMCs at the interface. This is determined with the process parameters such as the current of the TIG process. While the presence of IMC is essential to forming a joint, its high thickness degrades the joints. $TiAl_3$ and TiAl are the main phases that form during weld-brazing of Al to Ti [110]. Even a partial melting of Ti makes a complex microstructure in the FZ as well as a variety of IMCs at the interface [111].

For weld-brazing of dissimilar materials in lap configuration where the low-melting material is placed on top, it should be melted to flow on the surface of the bottom layer. When the low-melting material is placed on the bottom, the upper material should be melted only partially without reaching the lower material. The heating of the upper material is performed to an extent only to provide enough heat to melt the lower material. In fact, in this case the melting of the upper material is not necessary but is not inevitable.

To enhance the joint strength of dissimilar materials, the best way is to control the type of the IMC layer at the interface. One approach to do so is to choose a proper filler material to control the type of the IMC layers. As an example, in the case of weld-brazing of Al alloy to stainless steel, using Al—Cu filler material leads to better performance of the joint with respect to Al—Si filler material [112]. In the former, Al_8Fe_2Si and $(Al, Si)_{13}Fe_4$ are formed, which are brittle, while in the latter $Al_{13}(Fe, Cu)_4$ is formed, which has a higher crack resistance [112]. Another way for enhancing the joint strength is to control the thickness of the IMC layer. In the case of Al/galvanized steel, while preheating of the workpieces is needed to ensure a good wettability of the low-melting material over the other one, it should be kept as low as possible to avoid the growth of the IMCs [113]. CMT welding process has found its applications for weld-brazing of dissimilar materials to Al due to its controlled heat input [114].

While carbon steels can be galvanized prior to weld-brazing to Al, stainless steels cannot be galvanized due to their propensity to liquid metal embrittlement [115]. Therefore, other techniques are needed to produce a surface layer on stainless steel in order to improve the wettability of molten Al on its surface. Figure 2.33a shows the schematic of the stages of a process wherein the stainless steel is surfaced with Al by friction surfacing process. Then, a fusion welding process such as CMT is used to perform weld-brazing of Al to this surfaced stainless steel, as is observed in Figure 2.33b. Friction surfacing is a solid-state welding process by which a strong metallurgical bond is formed between the stainless steel and Al. Then, the adjoint Al sheet is placed over this surfaced area and melted by CMT. The melted Al spreads over this surface and melts the Al on the surface of stainless steel, making a fusion weld with this surface (Figure 2.33c). The fusion of this surface should not reach to its interface with stainless steel and hence the heat input needs to be controlled [115].

2.11 Simulation of the Arc Welding Processes

The main intentions from simulation of welding are calculation of the residual stresses and distortion as well as the thermal history during welding. While thicker plates are more prone to residual stresses, the thinner ones are more likely to develop distortion due to lower stiffness [116]. The calibration of the parameters of the simulation model requires only a few experiments that lower the cost and save

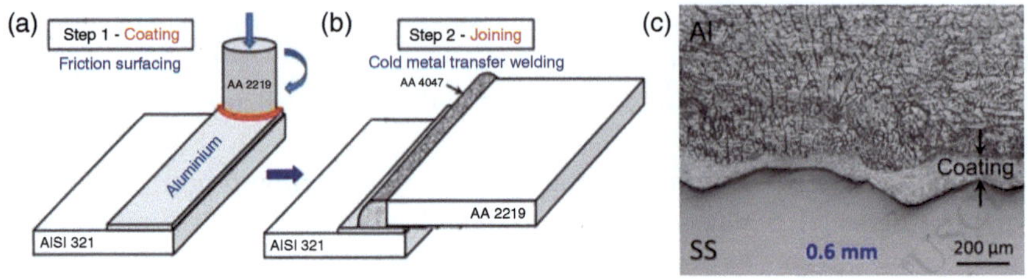

Figure 2.33 (a) Schematic of friction surfacing on stainless steel prior to (b) weld-brazing to Al. (c) Resulting microstructure at the interface. Source: Ref. [115]/with permission of Elsevier.

time. Additionally, simulation results give detailed information of distribution of residual stress and thermal history in every location of the weld, while experimental approaches yield this information only in specific zones. The simulation results also help to obtain a good understanding of the mechanism of the welding and development of residual stresses. The effect of process parameters and the joint geometry on the residual stress can be realized by simulation [117]. Residual stresses result from uneven heating and cooling in the material. At high temperatures, the yield strength of the material reduces, which facilitates plastic deformation induced by uneven thermal expansion and contraction in various locations. The residual stresses result from this uneven plastic deformation and remain in the workpiece after cooling down. At various temperatures, the mechanical properties of the material change, which should be taken into account during simulation. In addition, metallurgical changes may occur at high temperatures, which alter the properties of the material and can induce plastic deformation. These metallurgical changes add another complication to the simulation process. Al alloys encounter less complications for simulation due to absence of allotropic transformation in contrast to structural steels. So, less metallurgical complications exist for simulation of welding of Al alloys with respect to steels [118].

A key factor in thermal simulation of welding process is the shape of the heat source used in the model. Double ellipsoidal heat source is a most frequently used heat source for arc welding processes [108]. Modifications have been applied to the heat sources to simulate arc oscillation or pulsed arc [42]. For VP, separate models for the heat source when the electrode is negative and when it is positive are considered, as the heat delivery to the workpiece varies with polarity [119].

There is a method named the "contour method" used to calculate the residual stress without thermal analysis. In this method, the specimen is cut into two, leaving two free surfaces that are used to measure the height map. The average contour obtained from these two surfaces is then used to compute the residual stress through an elastic finite element (FE) model [120]. The basis of this procedure is relaxation of the residual stress on the cut surfaces that cause the materials to unload elastically. This method only yields the residual stress normal to the surface. Figure 2.34 illustrates the schematic of the contour method. Step A shows the distribution of the residual stress in the z-direction along the y-direction in the middle of the plate.

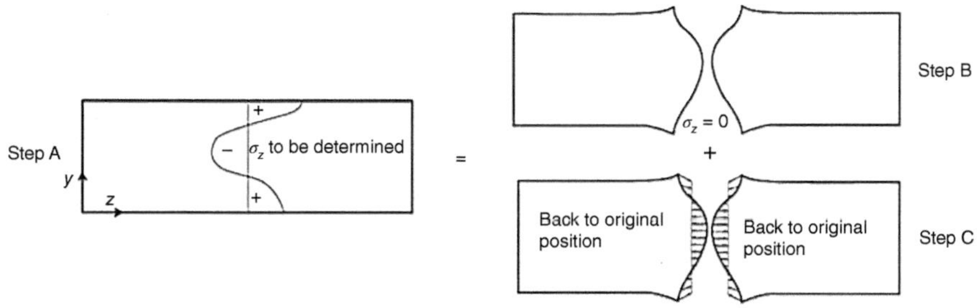

Figure 2.34 Schematic illustration of the contour method to calculate the residual stress. Source: Ref. [121]/John Wiley & Sons.

When the specimen is cut, the residual stress is relaxed causing an elastic deformation as shown in step B. The virtual stress required to force the deformed surfaces into their initial positions is measured through an elastic FE method, which is equivalent to the residual stress in the *z*-direction [121]. The accuracy of this method can be tested by comparing the obtained results with the ones obtained from XRD techniques [122]. An error exists in the data obtained from this method, as the displacement parallel to the cutting plane (the in-plane displacement) is unconstrained. This means that the nodes at the cutting surface do not return to their original position before cutting. However, this error is negligible and can be neglected [123].

By simulation, it is possible to investigate and understand the effect of multi-pass welding on the deformation and residual stress of the joint. Latter passes have higher stiffness due to constraints made by previous ones, and therefore lower deformation is developed. The latter passes have a heat-treating effect on the previous passes, which can release the residual stress [124]. Also, the most critical points in terms of maximum tensile residual stresses can be detected. This is visible in Figure 2.35, where a plate is welded to a sleeve by MIG welding in eight passes. The reduction in residual stress is visible when the latter passes are performed. Also, the highest residual stress is seen in the toe of the last pass, which is due to the higher stiffness when doing the last pass. Moreover, no heat-treating effect exists for the last pass, and hence, it possesses the highest residual stress [124].

This behavior is shown in another way during welding a V-groove of a thick Al plate in Figure 2.36a. The residual stresses as well as the temperature history at point *A* are presented in Figure 2.36b and c, respectively. After completion of the weld, the residual stress is almost zero due to the heat effects of the latter passes. The temperature evolution in this point shows how the peak temperature decreases, as the distance between the performed pass and point *A* increases.

Welding sequence is another factor that influences the residual stress and deformation. There are some basic rules to minimize distortion [116] which help to narrow the parameters window for welding. Despite this, obtaining the optimum parameters by experiment is still costly. Simulation provides this opportunity to investigate several approaches of welding sequence to obtain the optimum weld in terms of residual stress and distortion [126]. The restraint during welding is another important factor that determines the residual stress and distortion. Figure 2.37

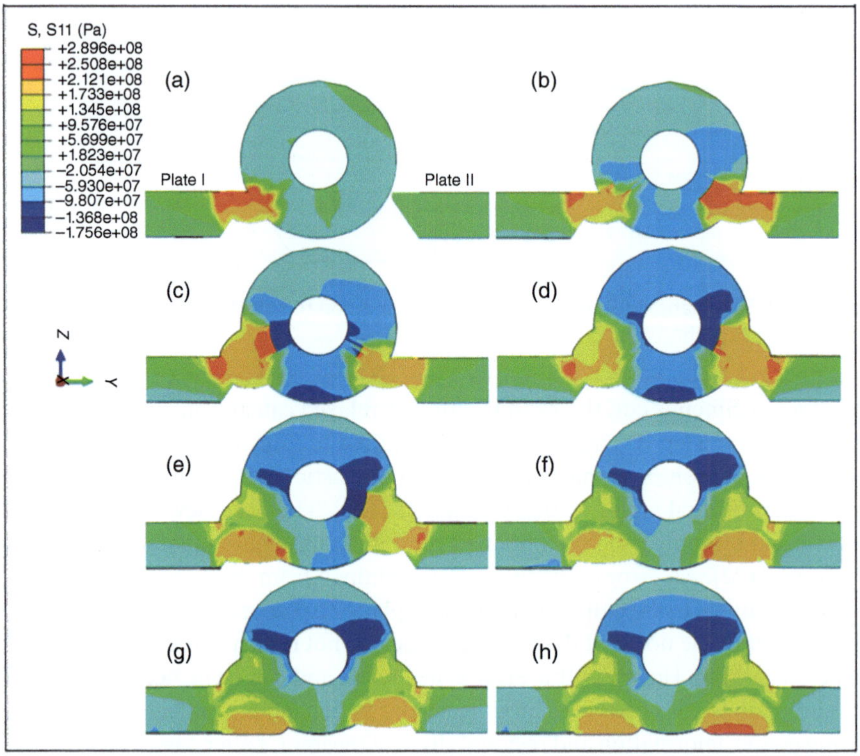

Figure 2.35 Contours of longitudinal residual stresses during welding a plate to a sleeve made of Al. (a–h) corresponds to the first pass to eighth pass. Source: Ref. [124]/Sage Publications.

shows the schematic of two Al sheets restrained by two vertical and horizontal forces. Various values of these forces yield various amounts of residual stress and distortion. A higher vertical force with zero horizontal force is the best practice to obtain a less concentrated residual stress and a smaller angular distortion [122].

Simulation also provides a tool to study the mechanism of droplet transfer in MIG welding [127]. Simulation is a strong tool to describe the phenomena that are not easy to explain by the physics of the process. For example, it is well known that the heat delivery from the electrode to the workpiece is higher when the electrode is negative. As mentioned before, in VP-TIG welding by increasing the percentage of positive polarity (%EP), the penetration increases. A simulation approach was used wherein the thermal efficiency for EN and EP periods was calibrated to attain a good match between the cross section of the simulated welds and the experiments [18]. The results showed that the thermal efficiency during EP is 0.9 while for EN is 0.25. This is due to the oxide cleaning effect, which reduces the needed voltage (Figure 2.3). The increased heat input is explained by the random walk of the cathode spot, which causes the more concentrated heat flux [18].

The common approach for numerical simulation is the thermal elastic–plastic method, which simulates the welding residual stress and distortion for small

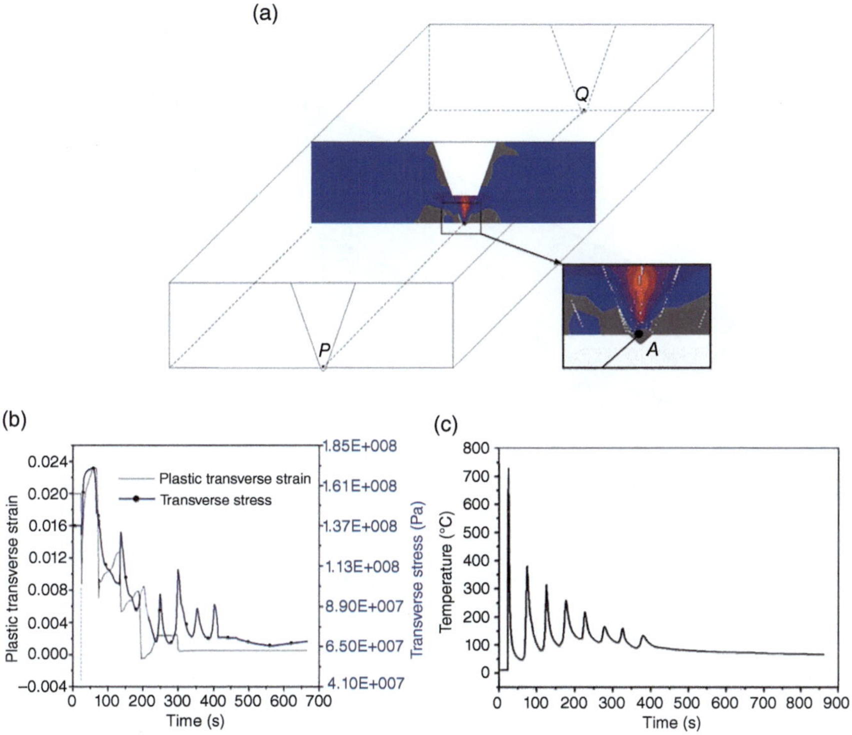

Figure 2.36 (a) A point at the bottom of a V-groove of a thick Al plate welded in several passes. (b) Plastic deformation and (c) thermal history at point *A*. Source: Ref. [125]/Springer Nature.

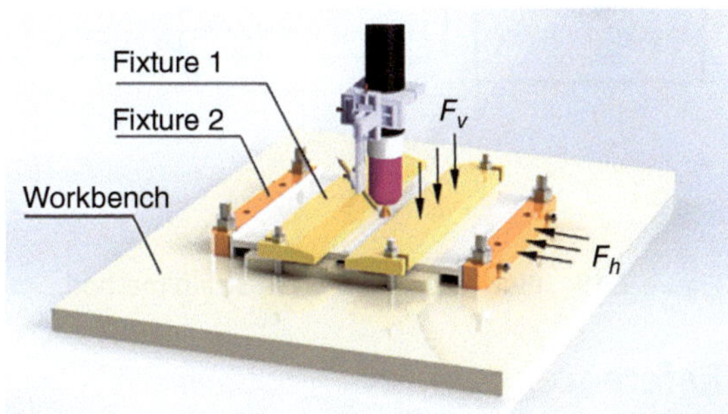

Figure 2.37 Schematic of the restraint applied on two Al sheets clamped by two vertical force (F_v) and horizontal force (F_h). Source: Ref. [122]/with permission of Elsevier.

structures. For large and complex thin-walled structures, the inherent strain method is used, which is based on the calculation of the residual stress and distortion by linear elastic method by knowing the inherent strain [128]. Figure 2.38 shows a train body based on a shell model which consists of several joints such as butt, lap, and Tee. Individual thermal elastic–plastic FE simulations are carried out for each joint, and then the obtained inherent deformations are used in an elastic FE simulation (Figure 2.39) [118].

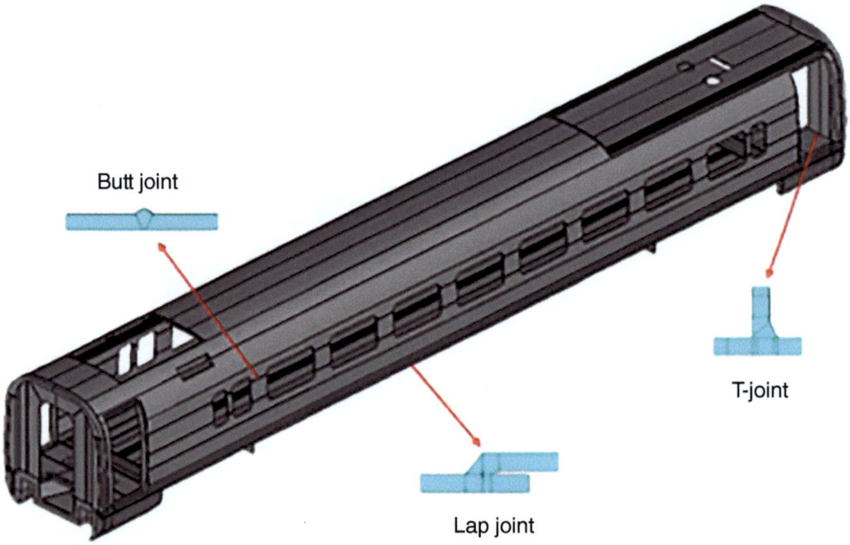

Figure 2.38 Shell model of a train body. Source: Ref. [128]/with permission of Elsevier.

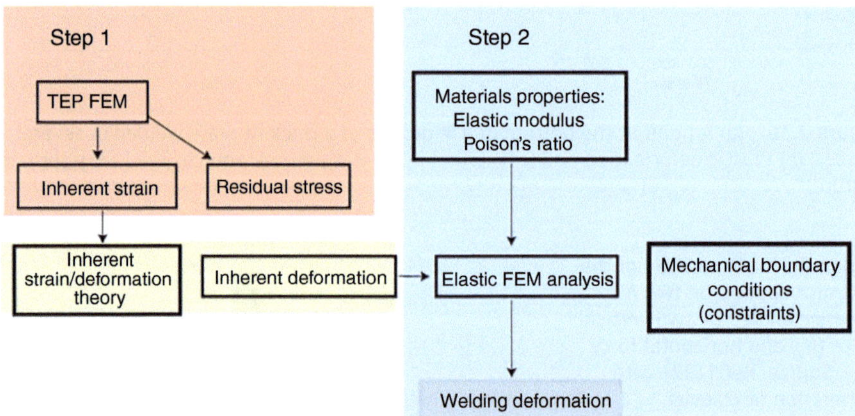

Figure 2.39 Flowchart of inherent strain method. Source: Ref. [118]/Springer Nature.

References

1 Doshi, S.J., Gohil, A.V., Mehta, N.D. et al. (2018). Challenges in fusion welding of Al alloy for body in white. *Materials Today: Proceedings* 5 (2): 6370–6375.

2 Jorge, V.L., Teixeira, F.R., Scotti, A. et al. (2023). The significance of supplementary shielding in WAAM of aluminium thin walls. *Journal of Manufacturing Processes* 106: 520–536.

3 Huang, L., Hua, X., Wu, D. et al. (2019). A study on the metallurgical and mechanical properties of a GMAW-welded Al-mg alloy with different plate thicknesses. *Journal of Manufacturing Processes* 37: 438–445.

4 Dilthey, U. (2005). Schweißen von Aluminiumwerkstoffen. In: *Schweißtechnische Fertigungsverfahren 2, Verhalten der Werkstoffe beim Schweißen* (ed. U. Dilthey), 217–237. Springer Verlag.

5 Wan, Z., Meng, D., Zhao, Y. et al. (2021). Improvement on the tensile properties of 2219-T8 aluminum alloy TIG welding joint with weld geometry optimization. *Journal of Manufacturing Processes* 67: 275–285.

6 Weber, M., Jeram, C., Schulmeister, K. et al. (2018). Erfassung und Bewertung der direkten und indirekten UV-Exposition bei unterschiedlichen Schweißverfahren. *NIR 2018: Wellen – Strahlung – Felder* 50: 75–85.

7 Mathers, G. (2002). *The Welding of Aluminium and its Alloys*. Elsevier.

8 Nölle, P. (2011). Verzugarmes Schweißen von Aluminiumwerkstoffen. *Lightweight-Design* 4 (2): 58–63.

9 Vasu, K., Chelladurai, H., Ramaswamy, A. et al. (2019). Effect of fusion welding processes on tensile properties of armor grade, high thickness, non-heat treatable aluminium alloy joints. *Defence Technology* 15 (3): 353–362.

10 Balanovskii, A.E. (2019). A new mechanism of interaction between a welding arc discharge of reverse-polarity direct current and an aluminum surface. *Teplofizika vysokikh temperatur* 57: 784–797.

11 Cirino, L.M. and Dutra, J.C. (2010). The influence of positive polarity time on GTAW AC of aluminium. *Welding International* 24 (11): 825–833.

12 Schellhase, M. (1985). *Der Schweißlichtbogen-ein technologisches Werkzeug*. Verlag Technik.

13 Siewert, E., Hussary, N., Schnick, M. et al. (2018). New GTAW variant for high-throughput aluminum welding. *Welding in the World* 62: 385–391.

14 Kutelu, B.J., Seidu, S.O., Eghabor, G.I. et al. (2018). Review of GTAW welding parameters. *Journal of Minerals and Materials Characterization and Engineering* 6 (05): 541.

15 Sathishkumar, D. and Das, A.D. (2021). Investigations on effect of process parameters on GTAW of aluminium alloy welding using full factorial design technique. *Materials Today Proceedings* 37: 621–626.

16 Sethuraman, S., Balaji, D., Sivaramakrishnan, V. et al. (2018). Optimization of GTAW process parameters on aluminum alloy 6063 on Taguchi method. *International Research Journal of Engineering and Technology* 5: 1129–1135.

17 Annamalai, D., Nampoothiri, J., Manikandan Rajam, P.K. et al. (2023). Optimization of ultrasonic-assisted TIG (UA-TIG) welding process parameters for AA7075 alloy joints using RSM-GA approach. *Journal of Testing and Evaluation* 51 (5): 3369–3389.

18 Jeong, H., Park, K., Baek, S. et al. (2019). Thermal efficiency decision of variable polarity aluminum arc welding through molten pool analysis. *International Journal of Heat and Mass Transfer* 138: 729–737.

19 Grinyuk, A., Korzhik, V.N., Shevchenko, V. et al. (2015). Main tendencies in development of plasma-arc welding of aluminium alloys. *The Paton Welding Journal* 11: 31–41.

20 González, L.R., Gonzalez, A.R., Pimentel, R.E. et al. (2024). MIG–VP: negative current effect and proportion of electrode negative on the penetration

aluminium weld bead. In: *Interconnections of Knowledge: Multidisciplinary Approaches* (ed. Sven Editora Academica), 811–831. Seven Publicações Ltda.

21 Padmanabham, G., Pandey, S., and Schaper, M. (2005). Pulsed gas metal arc welding of Al–cu–Li alloy. *Science and Technology of Welding and Joining* 10 (1): 67–75.

22 Gonzalez Olivares, E.A. (2022). Hybridization between plasma and MIG processes in a tandem configuration for aluminum joints. Dissertation. RWTH Aachen University.

23 Bai, Y., Gao, H.-M., and Qiu, L. (2010). Droplet transition for plasma-MIG welding on aluminium alloys. *Transactions of Nonferrous Metals Society of China* 20 (12): 2234–2239.

24 Lowke, J. (2011). Globula and spray transfer in MIG welding. *Welding in the World* 55: 19–23.

25 Wang, C., Zhang, H., Yan, Z. et al. (2021). Study on medium-thick Al-alloy T-joints by dual P-GMAW bilateral synchronous welding. *Metals* 11 (11): 1794.

26 Liu, Z., Fang, Y., Cui, S. et al. (2016). Stable keyhole welding process with K-TIG. *Journal of Materials Processing Technology* 238: 65–72.

27 Wang, Y., Qi, B., Cong, B.Q. et al. (2018). Keyhole welding of AA2219 aluminum alloy with double-pulsed variable polarity gas tungsten arc welding. *Journal of Manufacturing Processes* 34: 179–186.

28 Singh, S.R. and Khanna, P. (2021). A-TIG (activated flux tungsten inert gas) welding: – a review. *Materials Today Proceedings* 44: 808–820.

29 Hui, L. and Jia-sheng, Z. (2022). Study on removal mechanism of oxide film in direct current straight polarity tungsten arc welding of aluminum alloy. *Welding International* 36 (10): 589–596.

30 Li, H., Zou, J., Yao, J. et al. (2017). The effect of TIG welding techniques on microstructure, properties and porosity of the welded joint of 2219 aluminum alloy. *Journal of Alloys and Compounds* 727: 531–539.

31 Ahmadi, E., Ebrahimi, A., Hoseinzadeh, A. et al. (2020). Microstructure evolution and mechanical properties of 2219 aluminum alloy A–TIG welds. *The Physics of Metals and Metallography* 121: 483–488.

32 Wu, K., Liu, Z., Xie, P. et al. (2019). A comparative study on the bead profile and microstructural characteristics of aluminum alloy welds produced by single and double pulsed tandem gas metal arc welding. *Materials Research Express* 6 (8): p. 0865j1.

33 Miyazaki, H., Miyauchi, H., Sugiyama, Y. et al. (1993). Puckering in aluminium alloy welds-prevention using double wire MIG welding. *Welding International* 7 (6): 431–437.

34 Wu, K., Zhan, J., Cao, X. et al. (2021). Dynamic metal transfer behavior in double-wire DP-GMAW of aluminum alloy under different pulse phases. *Journal of Manufacturing Science and Engineering* 143 (4): 041002.

35 Garcia, R., Lopez, V., Bedolla, E. et al. (2002). MIG welding process with indirect electric arc. *Journal of Materials Science Letters* 21: 1965–1967.

36 Garcia, R., Lopez, V., Bedolla, E. et al. (2003). A comparative study of the MIG welding of Al/TiC composites using direct and indirect electric arc processes. *Journal of Materials Science* 38: 2771–2779.

37 Garcia, R., Lopez, V., and Bedolla, E. (2007). Welding of aluminium by the MIG process with indirect electric arc (MIG-IEA). *Journal of Materials Science* 42: 7956–7963.

38 García, R., López, V.H., Natividad, C. et al. (2011). Fusion welding with indirect electric arc. In: *Arc Welding* (ed. W. Sudnik), 21–44. IntechOpen.

39 Ambriz, R., Barrera, G., García, R. et al. (2009). A comparative study of the mechanical properties of 6061-T6 GMA welds obtained by the indirect electric arc (IEA) and the modified indirect electric arc (MIEA). *Materials & Design* 30 (7): 2446–2453.

40 Cornacchia, G. and Cecchel, S.J.M. (2020). Study and characterization of EN AW 6181/6082-T6 and EN AC 42100-T6 aluminum alloy welding of structural applications: metal inert gas (MIG), cold metal transfer (CMT), and fiber laser-MIG hybrid comparison. *Metals* 10 (4): 441.

41 Chen, J., Amirkhiz, B.S., Zhang, R. et al. (2020). On the joint formation and interfacial microstructure of cold metal transfer cycle step braze welding of aluminum to steel butt joint. *Metallurgical and Materials Transactions A* 51 (10): 5198–5212.

42 Zapico, E.P., Lutey, A.H.A., Ascari, A. et al. (2018). An improved model for cold metal transfer welding of aluminium alloys. *Journal of Thermal Analysis and Calorimetry* 131: 3003–3009.

43 Guojin, L., Peilei, Z., Xi, W. et al. (2018). Gap bridging of 6061 aluminum alloy joints welded by variable-polarity cold metal transfer. *Journal of Materials Processing Technology* 255: 927–935.

44 Zhang, P., Li, G., Yan, H. et al. (2020). Effect of positive/negative electrode ratio on cold metal transfer welding of 6061 aluminum alloy. *The International Journal of Advanced Manufacturing Technology* 106: 1453–1464.

45 Selvi, S., Vishvaksenan, A., and Rajasekar, E. (2018). Cold metal transfer (CMT) technology – an overview. *Defence Technology* 14 (1): 28–44.

46 An'Shakov, A., Urbakh, A.E., Kobeleva, E.N. et al. (2004). Development of a tungsten cathode for microplasma welding of components with reversed polarity current. *Welding International* 18 (8): 641–645.

47 Treutler, K., Brechelt, S., Wiche, H. et al. (2021). Beneficial use of hyperbaric process conditions for welding of aluminium and copper alloys. *Welding in the World* 65 (8): 1623–1631.

48 Sarpe, M., K. Treutler, V. Wesling et al. (2023). Einfluss hyperbarer Prozessbedingungen auf die Porenbildung beim MSG-Schweißen von verzinkten höherfesten Fahrwerkstählen. *Tagungsband 5. Niedersächsisches Symposium Materialtechnik*, Clausthal-Zellerfeld (23–24 February 2023).

49 Grinyuk, A., Коржик, В.Н., Шевченко, В.Е. et al. (2016). Hybrid technologies of welding aluminium alloys based on consumable electrode arc and constricted arc. *Avtomaticheskaya Svarka* 5 (6): 98–103.

50 Babych, O., Korzhyk, V., and Grynyuk, A. (2020). Hybrid welding of aluminium 1561 and 5083 alloys using plasma-arc and consumable electrode arc (plasma-MIG). *Avtomatičeskaâ Svarka (Kiev)* 7: 11–22.

51 Guo, Y., Pan, H., Ren, L. et al. (2018). An investigation on plasma-MIG hybrid welding of 5083 aluminum alloy. *The International Journal of Advanced Manufacturing Technology* 98: 1433–1440.

52 Yuji, T., Tashiro, S., Kinoshita, H. et al. (2024). An investigation on plasma-MIG hybrid welding process of thick plate aluminum. *Journal of Advanced Joining Processes* 9: 100188.

53 Lee, H.-k., Park, S.-h., and Kang, C.-Y. (2015). Effect of plasma current on surface defects of plasma-MIG welding in cryogenic aluminum alloys. *Journal of Materials Processing Technology* 223: 203–215.

54 Yang, T., Chen, L., Zhuang, Y. et al. (2020). Arcs interaction mechanism in plasma-MIG hybrid welding of 2219 aluminium alloy. *Journal of Manufacturing Processes* 56: 635–642.

55 Lee, H.-K., Chun, K.-S., Park, S.-h. et al. (2015). Control of surface defects on plasma-MIG hybrid welds in cryogenic aluminum alloys. *International Journal of Naval Architecture and Ocean Engineering* 7 (4): 770–783.

56 Lin, H.-L., Chen, G.-C., and Cheng, C.-M. (2022). Effects of plasma-MIG hybrid welding parameters on performance and hot cracking susceptibility of 5083 Al alloy butt-joint welds. *The International Journal of Advanced Manufacturing Technology* 122 (9): 4055–4064.

57 Ono, K., Liu, Z., Era, T. et al. (2009). Development of a plasma MIG welding system for aluminium. *Welding International* 23 (11): 805–809.

58 Mamat, S.B., Tashiro, S., Tanaka, M. et al. (2018). Study on factors affecting the droplet temperature in plasma MIG welding process. *Journal of Physics D: Applied Physics* 51 (13): 135206.

59 Yan, B., Gao, H.-m., Wu, L. et al. (2010). Influence of plasma-MIG welding parameters on aluminum weld porosity by orthogonal test. *Transactions of Nonferrous Metals Society of China* 20 (8): 1392–1396.

60 Kanemaru, S., Sasaki, T., Sato, T. et al. (2015). Study for the mechanism of TIG-MIG hybrid welding process. *Welding in the World* 59: 261–268.

61 Huang, J., Chen, H., He, J. et al. (2019). Narrow gap applications of swing TIG-MIG hybrid weldings. *Journal of Materials Processing Technology* 271: 609–614.

62 Ye, Z., Huang, J., Cheng, Z. et al. (2018). Study on butt joining 5052 aluminum alloy/Q235 mild steel by MIG-TIG double-sided arc welding-brazing process. *Welding in the World* 62: 145–154.

63 Peasura, P. and Watanapa, A. (2012). Influence of shielding gas on aluminum alloy 5083 in gas tungsten arc welding. *Procedia Engineering* 29: 2465–2469.

64 Guo, C., Zhang, H., Wu, Z. et al. (2019). Effects of ag on the age hardening response and intergranular corrosion resistance of Al–mg alloys. *Materials Characterization* 147: 84–92.

65 Ebrahimzadeh, H., Farhangi, H., and Mousavi, S.A.A.A. (2020). Hot cracking in autogenous welding of 6061-T6 aluminum alloy by rectangular pulsed Nd: YAG laser beam. *Welding in the World* 64: 1077–1088.

66 Abdollahi, A., Nganbe, M., and Kabir, A.S. (2024). An adapted approach for solidification crack elimination in Al7075 TIG welding. *Materials Research Express* 11 (4): 046509.

67 Fouad, R.A., Habba, M.I.A., Gamal, Y. et al. (2024). The influence of various welding wires on microstructure, and mechanical characteristics of AA7075 Al-alloy welded by TIG process. *Scientific Reports* 14 (1): 19023.

68 Bodaghi, F., Movahedi, M., Na, S.-J. et al. (2024). Effect of welding current and speed on solidification cracking susceptibility in gas tungsten arc fillet welding of dissimilar aluminum alloys: coupling a weld simulation and a cracking criterion. *Journal of Materials Research and Technology* 30: 4777–4785.

69 Mercan, E., Ayan, Y., Kahraman, N. et al. (2020). Investigation on joint properties of AA5754 and AA6013 dissimilar aluminum alloys welded using automatic GMAW. *Engineering Science and Technology an International Journal* 23 (4): 723–731.

70 Yadav, A.K., Agrawal, M.K., Saxena, K.K. et al. (2024). Effect of GTAW process parameters on weld characteristics and microstructural studies of dissimilar welded joints of AA5083 and AA6082: optimization technique. *International Journal on Interactive Design and Manufacturing* 18 (3): 1151–1160.

71 Rojas, H., Molina-Ocampo, A., Valdez, S. et al. (2020). The impact of heat input on the microstructures, fatigue behaviors, and stress lives of TIG-welded 6061-T6 alloy joints. *Materials Research Express* 7 (12): 126512.

72 Bindu, A.H., Chaitanya, B.s.K., Ajay, K. et al. (2020). Investigation on feasibility of dissimilar welding of AA2124 and AA7075 aluminium alloy using tungsten inert gas welding. *Materials Today Proceedings* 26: 2283–2288.

73 Baskoro, A.S., Amat, M.A., Pratama, A.I. et al. (2019). Effects of tungsten inert gas (TIG) welding parameters on macrostructure, microstructure, and mechanical properties of AA6063-T5 using the controlled intermittent wire feeding method. *The International Journal of Advanced Manufacturing Technology* 105: 2237–2251.

74 Yan, Z., Yuan, T., Chen, S.J. et al. (2019). Microstructural refinement of 6061 and 5052 aluminium alloys by arc oscillation. *Materials Science and Technology* 35 (13): 1651–1655.

75 Bunn, A., Evans, P., Bristow, D.J. et al. (1998). Modelling of the effectiveness of Al–Ti–B refiners in commercial purity aluminium. In: *Essential Readings in Light Metals: Volume 3 Cast Shop for Aluminum Production* (ed. J.F. Grandfield and D.G. Eskin), 387–392. Springer.

76 Tang, Z., Schempp, P., Seefeld, T. et al. (2011). Kornfeinung beim WIG-und Laserstrahlschweißen von Aluminiumlegierungen. In: *DVS-Berichte*, 153–163. DVS-Verlag.

77 Schempp, P., Pittner, A., and Rethmeier, M. (2014). Kornstruktur in Aluminium-WIG-Schweißnähten. *Schweißen und Schneiden: Fachzeitschrift für die Schweiss und verwandte Verfahren* 66 (3): 120–125.

78 Easton, M. and St John, D. (1999). Grain refinement of aluminum alloys: part I. The nucleant and solute paradigms – a review of the literature. *Metallurgical and Materials Transactions A* 30: 1613–1623.

79 Easton, M. and St John, D. (1999). Grain refinement of aluminum alloys: part II. Confirmation of, and a mechanism for, the solute paradigm. *Metallurgical and Materials Transactions A* 30: 1625–1633.

80 Chen, Q.-H., Lin, S.-B., Yang, C.-L. et al. (2016). Effect of ultrasound on heterogeneous nucleation in TIG welding of Al–Li alloy. *Acta Metallurgica Sinica* 29: 1081–1088.

81 Chen, Q.-H., Lin, S., Yang, C.-L. et al. (2016). Effect of ultrasonic impact on the microstructure of welded joint of 2195 Al–Li alloy. *Acta Metallurgica Sinica* 29: 367–372.

82 Chen, Q., Lin, S., Yang, C. et al. (2017). Grain fragmentation in ultrasonic-assisted TIG weld of pure aluminum. *Ultrasonics Sonochemistry* 39: 403–413.

83 Li, Y., Zou, W., Lee, B. et al. (2020). Research progress of aluminum alloy welding technology. *The International Journal of Advanced Manufacturing Technology* 109: 1207–1218.

84 Sharma, P., Chattopadhyaya, S., Kumar Singh, N. et al. (2022). Recent developments in the design, development, and analysis of the influence of external magnetic-field on gas-metal arc welding of non-ferrous alloys: review on optimization of arc-structure to enhance the morphology, and mechanical properties of welded joints for automotive applications. *Heliyon* 8 (12): e11812.

85 Verma, R.P., Pandey, K.N., Kovács, A. et al. (2023). Difficulties and redressal in joining of aluminium alloys by GMA and GTA welding: a review. *Journal of Materials Research and Technology* 23: 2576–2586.

86 Wang, L., Suo, Y., Wu, C. et al. (2018). Effect of pulse frequency on microstructure and mechanical properties of 2198 Al–Li alloy joints obtained by ultrahigh-frequency pulse AC CMT welding. *Materials (Basel)* 12 (1): 79.

87 Wang, Y., Li, H., Li, Z. et al. (2023). Refining microstructure of medium-thick AA2219 aluminium alloy welded joint by ultrasonic frequency double-pulsed arc. *Journal of Materials Research and Technology* 23: 3048–3061.

88 Bi, J., Liu, Z., Chi, J. et al. (2023). Formation mechanisms and control strategies of FQZ softening in Al–Li alloy welded joint. *Journal of Materials Research and Technology* 23: 2810–2823.

89 Hu, Y., Wu, S.C., and Chen, L. (2019). Review on failure behaviors of fusion welded high-strength Al alloys due to fine equiaxed zone. *Engineering Fracture Mechanics* 208: 45–71.

90 Zhang, D., Aiping, W., Zhao, Y. et al. (2021). Effects of the number of welding passes on microstructure and properties of 2219-C10S aluminum alloy TIG-welded joints. *Journal of Materials Engineering and Performance* 30: 3537–3546.

91 Zhang, D.-k., Zhao, Y., Dong, M.-y. et al. (2019). Effects of weld penetration on tensile properties of 2219 aluminum alloy TIG-welded joints. *Transactions of Nonferrous Metals Society of China* 29 (6): 1161–1168.

92 Zhang, D., Wang, G., Wu, A. et al. (2019). Study on the inconsistency in mechanical properties of 2219 aluminium alloy TIG-welded joints. *Journal of Alloys and Compounds* 777: 1044–1053.

93 Wan, Z., Zhao, Y., Wang, Q. et al. (2022). Microstructure-based modeling of the PMZ mechanical properties in 2219-T8 aluminum alloy TIG welding joint. *Materials & Design* 223: 111133.

94 Sugiyama, Y., Nakata, J., and Miyauchi, H. (1993). Reducing smut in aluminium alloy welds using double wire MIG (DWM) welding. *Welding International* 7 (3): 177–182.

95 Samiuddin, M., Li, J.l., Taimoor, M. et al. (2021). Investigation on the process parameters of TIG-welded aluminum alloy through mechanical and microstructural characterization. *Defence Technology* 17 (4): 1234–1248.

96 Zhu, C., Cheon, J., Tang, X. et al. (2018). Molten pool behaviors and their influences on welding defects in narrow gap GMAW of 5083 Al-alloy. *International Journal of Heat and Mass Transfer* 126: 1206–1221.

97 Koli, Y., Yuvaraj, N., Aravindan, S. et al. (2021). Enhancement of mechanical properties of 6061/6082 dissimilar aluminium alloys through ultrasonic-assisted cold metal transfer welding. *The Arabian Journal for Science and Engineering* 46 (12): 12089–12104.

98 Kou, S. (2003). Solidification and liquation cracking issues in welding. *JOM: The Journal of the Minerals, Metals & Materials Society* 55: 37–42.

99 Sokoluk, M., Cao, C., Pan, S. et al. (2019). Nanoparticle-enabled phase control for arc welding of unweldable aluminum alloy 7075. *Nature Communications* 10 (1): 98.

100 Murali, N., Sokoluk, M., and Li, X. (2021). Study on aluminum alloy joints welded with nano-treated Al–mg–Mn filler wire. *Materials Letters* 283: 128739.

101 Wan, Z., Wang, Q., Zhao, Y. et al. (2022). Improvement in tensile properties of 2219-T8 aluminum alloy TIG welding joint by PMZ local properties and stress distribution. *Materials Science and Engineering: A* 839: 142863.

102 Kaliyaperumal, B., Soundararajan, R., Rengaraj, J. et al. (2021). An experimental investigation on hot cracking and properties of TIG welded AA2014 alloy using continuous and pulsed current. SAE Technical Paper.

103 Yang, Y.P. (2021). Recent advances in the prediction of weld residual stress and distortion – part 2. *Welding Journal* 100 (6): 193S–206S.

104 Yang, Y., Dong, P., Zhang, J. et al. (2000). A hot-cracking mitigation technique for welding high-strength aluminum alloy. *Welding Journal* 79 (1): 9.

105 Kang, M. and Kim, C. (2015). Joining Al 5052 alloy to aluminized steel sheet using cold metal transfer process. *Materials & Design* 81: 95–103.

106 Alikhani, A., Beygi, R., Mehrizi, M.Z. et al. (2021). Effect of mg and Si on intermetallic formation and fracture behavior of pure aluminum-galvanized carbon-steel joints made by weld-brazing. *Journal of Central South Univedrsity* 28 (11): 3626–3638.

107 Fan, D., Yang, N., Huang, J. et al. (2021). Plasma arc welding-brazing of aluminum to copper with SiO_2 nanoparticles strengthening. *Journal of Manufacturing Processes* 69: 253–260.

108 Huang, J., Liu, Y., Liu, S. et al. (2021). Process of welding-brazing and interface analysis of lap joint Ti–6Al–4V and aluminum by plasma arc welding. *Journal of Manufacturing Processes* 61: 396–407.

109 Singh, J., Arora, K.S., and Shukla, D.K. (2020). Lap weld-brazing of aluminium to steel using novel cold metal transfer process. *Journal of Materials Processing Technology* 283: 116728.

110 Zhang, Y., Huang, J., Ye, Z. et al. (2018). Influence of welding parameters on the IMCs and the mechanical properties of Ti/Al butt joints welded by MIG/TIG double-sided arc welding-brazing. *Journal of Alloys and Compounds* 747: 764–771.

111 Wei, S., Li, Y., Wang, J. et al. (2014). Use of welding–brazing technology on microstructural development of titanium/aluminum dissimilar joints. *Materials and Manufacturing Processes* 29 (8): 961–968.

112 Song, J., Lin, S., Yang, C.L. et al. (2010). Analysis of intermetallic layer in dissimilar TIG welding–brazing butt joint of aluminium alloy to stainless steel. *Science and Technology of Welding & Joining* 15 (3): 213–218.

113 Ma, H., Qin, G., Bai, X. et al. (2016). Effect of initial temperature on joint of aluminum alloy to galvanized steel welded by MIG arc brazing-fusion welding process. *The International Journal of Advanced Manufacturing Technology* 86: 3135–3143.

114 Burhanuddin, N., Redzuan, N., Ahmad, N. et al. (2020). Brief review on dissimilar welding using cold metal transfer. *IOP Conference Series: Materials Science and Engineering*, Putrajaya, Malaysia (8–9 December 2019). IOP Publishing.

115 Babu, S., Panigrahi, S.K., Janaki Ram, G.D. et al. (2019). Cold metal transfer welding of aluminium alloy AA 2219 to austenitic stainless steel AISI 321. *Journal of Materials Processing Technology* 266: 155–164.

116 Conrardy, C., Huang, T.D., Harwig, D. et al. (2006). Practical welding techniques to minimize distortion in lightweight ship structures. *Journal of Ship Production and Design* 22 (04): 239–247.

117 Hossain, I., Bhowmik, A., Pattanaik, A. et al. (2024). Computational investigation of plasma arc welding process for aluminium alloys. *Engineering Research Express* 6 (2): 025541.

118 Beygi, R., Marques, E., and da Silva, L.F. (2022). *Computational Concepts in Simulation of Welding Processes*. Springer.

119 Sun, Z., Han, Y., Du, M. et al. (2018). Numerical simulation of VPPA-GMAW hybrid welding of thick aluminum alloy plates considering variable heat input and droplet kinetic energy. *Journal of Manufacturing Processes* 34: 688–696.

120 Prime, M.B. and DeWald, A.T. The contour method. In practical residual stress measurement methods ed. G.S. Schajer, 2013: p. 109–138. *Wiley*.

121 Hosseinzadeh, F., Kowal, J., and Bouchard, P.J. (2014). Towards good practice guidelines for the contour method of residual stress measurement. *The Journal of Engineering* 2014 (8): 453–468.

122 Wang, Q., Zhao, Y., Zhao, T. et al. (2021). Influence of restraint conditions on residual stress and distortion of 2219-T8 aluminum alloy TIG welded joints based on contour method. *Journal of Manufacturing Processes* 68: 796–806.

123 Lei, Z., Zou, J., Wang, D. et al. (2020). Finite-element inverse analysis of residual stress for laser welding based on a contour method. *Optics & Laser Technology* 129: 106289.

124 Zhao, S., Li, Y., and You, H. (2023). Numerical simulation of welding deformation and residual stress in aluminum alloy plate-to-sleeve welded joints. *Proceedings of the Institution of Mechanical Engineers Part L Journal of Materials Design and Applications* 237 (1): 198–217.

125 Shu, F., Wu, L., Liu, Y. et al. (2019). Numerical simulation of residual stresses in multipass gas metal arc-welded joints of thick aluminum alloy plates. *Journal of Materials Engineering and Performance* 28: 3361–3371.

126 Khoshroyan, A. and Darvazi, A.R. (2020). Effects of welding parameters and welding sequence on residual stress and distortion in Al6061-T6 aluminum alloy for T-shaped welded joint. *Transactions of Nonferrous Metals Society of China* 30 (1): 76–89.

127 Ikram, A. and Chung, H. (2021). Numerical simulation of arc, metal transfer and its impingement on weld pool in variable polarity gas metal arc welding. *Journal of Manufacturing Processes* 64: 1529–1543.

128 Lu, Y., Zhu, S., Zhao, Z. et al. (2020). Numerical simulation of residual stresses in aluminum alloy welded joints. *Journal of Manufacturing Processes* 50: 380–393.

3

Beam Welding Processes

3.1 Introduction

Conventional arc welding processes to join thick Al plates need a high groove preparation, necessitating a multi-pass welding with a high heat input. Laser welding possesses a high energy density and is able to weld thick Al sheets in single pass with a lower heat input [1]. Laser welding of Al has found a high demand for electrical and structural components, especially in electric vehicles. It is used to join Al to other materials in the batteries of electric cars, for example the Al busbar and the steel pole of the battery [2]. Another application of LBW in electric cars is joining of the heat exchangers to the base plate of a battery tray [3].

Laser welding is also capable of joining Al to other materials in which the structural properties are of utmost importance. Examples include Al/Ti joints in the wings of airplanes where Ti as the crust is joined to Al as the honeycomb [4] and the passenger seat track [5] as shown in Figure 3.1.

For more than 15 years, Laser welding has been substituting riveting in joining of Al in the aerospace industry [6]. A potential usage of LBW is joining of stiffeners to the skin in the hull structure of the airplane (Figure 3.2a and b). This process is a good alternative for riveting (Figure 3.2c and d), because riveting needs an overlap, which adds to the weight. Furthermore, riveting has several steps, which makes it slow [6].

The joint configuration in this state is Tee. Two laser beams are used simultaneously on both sides to ensure that the vapor capillaries from both sides meet each other (Figure 3.3). In this case, the angle of the beams, their symmetry, and their energy need to be adjusted, so that the keyholes are stable and the skin will not be melted through. The melt-through of the skin causes aesthetic problems as well as problems in the aerodynamics. The high speed of welding and the small area of the weld zone minimize the heat input, and therefore the angle distortion and transverse distortion will be lessened. The hull structure of the airplane consists of several welds, and therefore to attain minimum distortion, the weld sequence needs to be planned carefully [6].

According to DIN 8593, laser welding is classified as a fusion welding process. This process is used for welding of materials with high thermal conductivity

Welding Metallurgy of Aluminium Alloys: Design, Processes, and Simulations, First Edition.
Reza Beygi, Eduardo André de Sousa Marques, Ricardo João Camilo Carbas, and Lucas Filipe Martins da Silva.

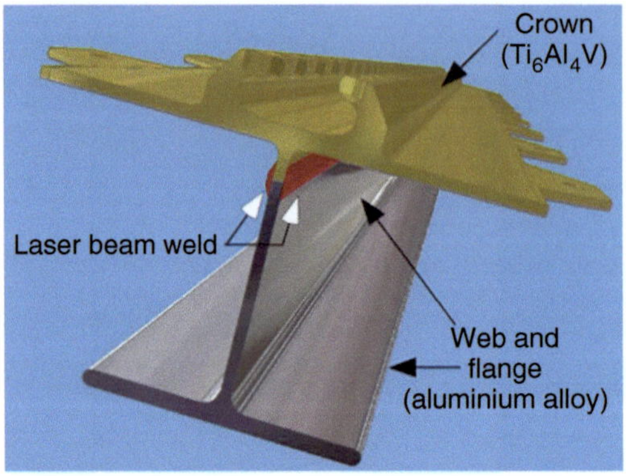

Figure 3.1 Titanium alloy and aluminum joined by laser welding in the passenger seat track. Source: Ref. [5].

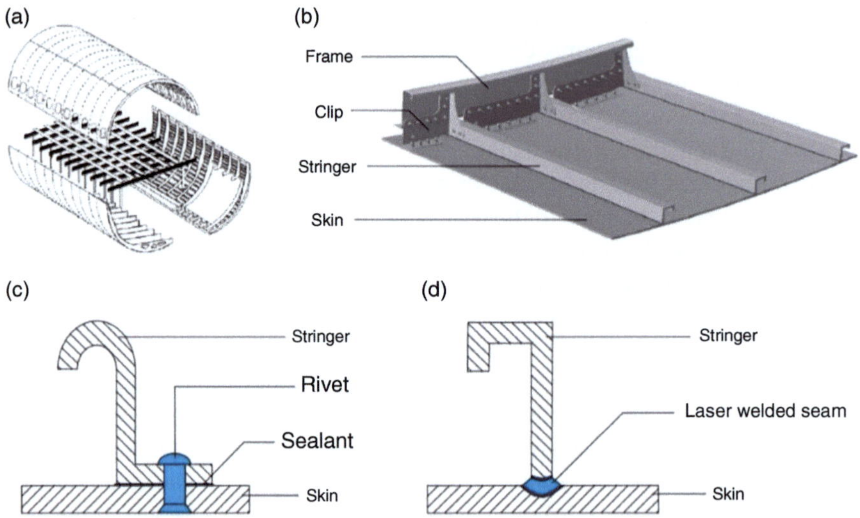

Figure 3.2 (a) Hull structure of the airplane. (b) Skin-stringer panel. (c) Riveting of the stringer to the skin. (d) Laser welding of the stringer to the skin. Source: Ref. [6].

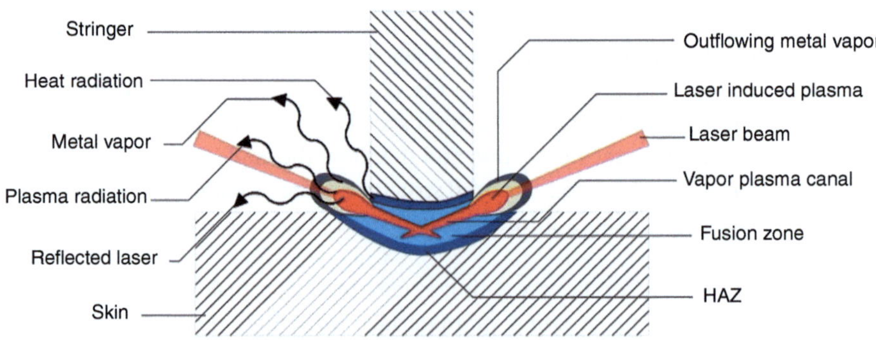

Figure 3.3 Laser beam welding of a Tee-joint. Source: Ref. [6].

due to its concentrated energy, whose power intensity can be controlled according to the application and purpose [7]. A narrow HAZ is one important characteristic of this process. During laser welding, defects such as depression and collapse are likely to occur due to high fluidity of the molten pool. Using the proper filler wire may help to eliminate these defects [8]. In laser welding, filler material is also used to prevent hot cracking. Filler material is more important when considering the high welding speed of this process, which promotes hot cracking. However, for thin sheets normally no filler material is used, as it necessitates a tedious optimization of the process [9]. Laser welding is very intolerable to the gap and is also prone to the pore formation when the weld penetration is deep [10].

During laser welding, a keyhole may be formed at high energy densities (higher than 10^6 W/cm^2) that cause a high penetration depth, making it suitable for welding thick sheets [11]. The keyhole is a result of vapor pressure that causes a high process dynamics accompanied by defects such as pores and spatters [12]. In fact, this vapor pressure opposes the surface tension that tries to close the hole [13]. Multiple reflections inside the keyhole cause a high absorption of the laser beam. The higher the ratio of the depth to the diameter, the higher the absorption of the laser beam, which is attributed to the multi-reflection of the beam [14]. When the energy distribution is lower than 10^6 W/cm^2, no keyhole is produced, making it suitable to join thin sheets (lower than 1 mm). This is referred to as the heat conduction mode laser welding. In this state, the mechanism of weld pool formation is like the one produced by TIG welding. Figure 3.4a and b shows two modes of laser welding.

Besides the keyhole, another form of cavity in the shape of a catenoid may form during laser welding of thin sheets referred to as the "laser donut welding" [15]. This catenoid-shaped cavity forms when the width of the melt pool is higher than 1.5 times the height of the melt pool. The reason that this cavity forms during laser welding of thin sheets is that the minimum surface area of the melt pool is achieved when a cavity with the shape of a catenoid is formed. In this case, the effect of vapor pressure is negligible due to the high opening of the hole [15].

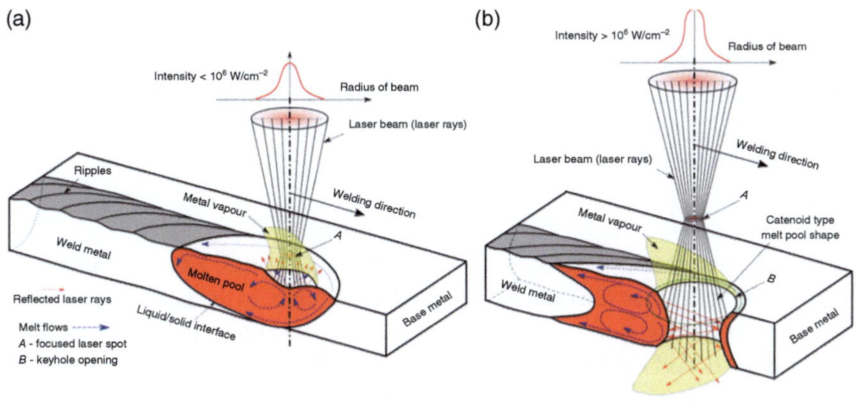

Figure 3.4 (a) Heat conduction mode laser welding and (b) keyhole laser welding. Source: Ref. [11].

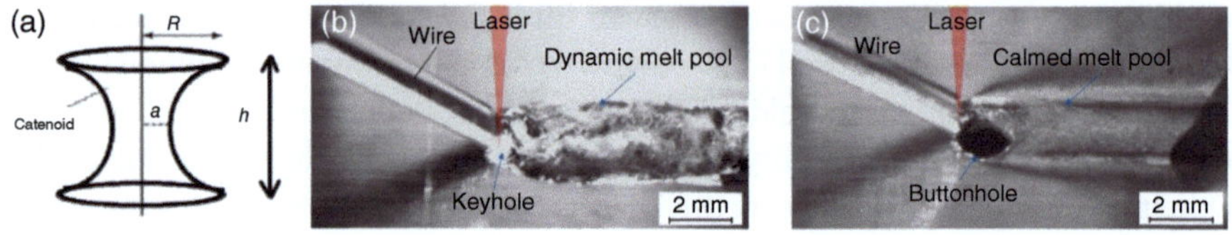

Figure 3.5 (a) Shape of the catenoid and the parameters that define its shape. Weld crown of a thin aluminum sheet welded (b) without and (c) with this catenoid hole. Source: Ref. [16]/MDPI/CC BY 4.0.

The shape of a catenoid and its dimensions is depicted in Figure 3.5a. The laser beam is in front of the catenoid causing a keyhole to form beside this catenoid. The melt in front of the catenoid is transported around the catenoid to its rear tranquilly without any spatter, which causes the surface appearance to be very smooth. Figure 3.5b and c shows the weld crown of a thin Al sheet welded without and with this catenoid hole. The main disadvantage of this method is the hole that remains at the end of the weld, and therefore, a run-off plate is needed to locate this hole away from the main seam of the weld. Also, a run-on plate is needed to form this hole at the beginning of the weld [16]. This is because this hole needs several millimeters at the beginning of the weld to be established [15].

Al is susceptible to pore formation when the welding speed during laser welding is low [17]. The pores form at the bottom of the keyhole. The standard ISO 13919-2 classifies the quality of the weld with respect to the pores in *B*, *C*, *D*, and non-acceptable with respect to the pores dimensions and their area fraction [1]. Various alloying elements of Al influence the power of the laser beam process that is needed to form a deep penetration as well as to have a stable keyhole. They also influence the viscosity of the melt and therefore the shape of the weld [6].

A high reflectivity of the laser beam in the wavelength range of 1030–1070 nm and 10,600 nm by Al causes a high energy consumption and thus needs a high investment [18]. The minimum beam irradiance required to melt the surface of a material is named the threshold beam irradiance (l_{th}) and is given by [19]:

$$l_{th} = k(T_m - T_0)/(AdJ_{\max}) \tag{3.1}$$

where k is the thermal conductivity, T_m is the melting temperature of the material, T_0 is the ambient temperature, A is the absorption of the surface, and d is the laser beam diameter. $J_{\max}$ is the degree of the interaction of the laser beam and material and is obtained from [19]:

$$J_{\max} = f\left(\frac{\alpha}{vd}\right) \tag{3.2}$$

where α is the thermal diffusivity and v is the welding speed. Laser is mainly at infrared wavelength (1064 nm) and Al has a high reflectance to the light in sub-band. Al has a high absorption rate for the blue light (450 nm) and green light (515 nm) [2]. Due to the reflection of the beam from the surface of Al, the equipment may be damaged by the reflection of the light, and therefore, a tilt angle can be considered to overcome this issue [20, 21]. Sandblasting is another method that can improve the laser absorption of Al due to surface roughness [22].

3.2 Laser Systems

The beam quality of a laser system is defined by a parameter named beam parameter product (BPP), which is defined by two parameters: the beam waist radius and the divergence angle (Figure 3.6a) [23]. The higher the BPP, the lower the beam quality. A comparison of different laser types is provided in Figure 3.6b. Disk laser and fiber laser are more attractive to the market as their BPP at high power is not very high. Furthermore, due to their development in recent years, their prices are more attractive [24]. Figure 3.6b presents the BPP versus the power for some laser systems.

By diode laser, the welding pool is very quiet and the spatter is very low. Therefore, the impurities are lower and the weld bead is uniform and clean. The profile of its energy intensity distribution is top-hat shaped [25]. The Gaussian profile has a redundant energy higher than the threshold needed for the process. It also has a smaller energy region than the threshold energy (Figure 3.7). Thus, the energy efficiency of diode laser welding is higher than the other laser welding techniques. In comparison with Gaussian profile, a steeper change of profile leads to a smaller HAZ. The lower wavelength of this process is in the maximum absorption of Al. That means lower laser is reflected back from the surface of Al and so lower energy is needed.

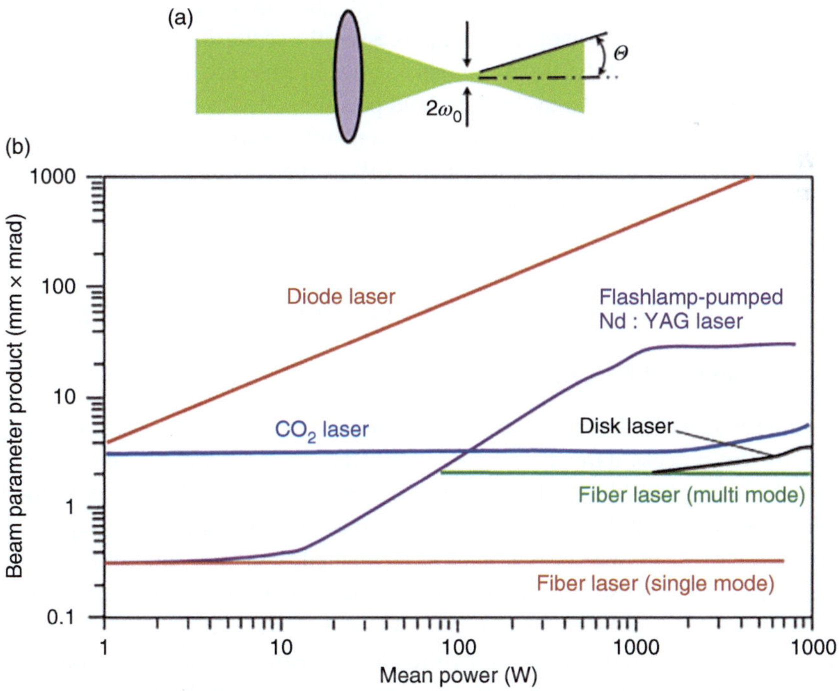

Figure 3.6 (a) Schematic of laser beam parameters and (b) a comparison of laser types with regard to the dependency of their beam parameter product and the mean power [24].

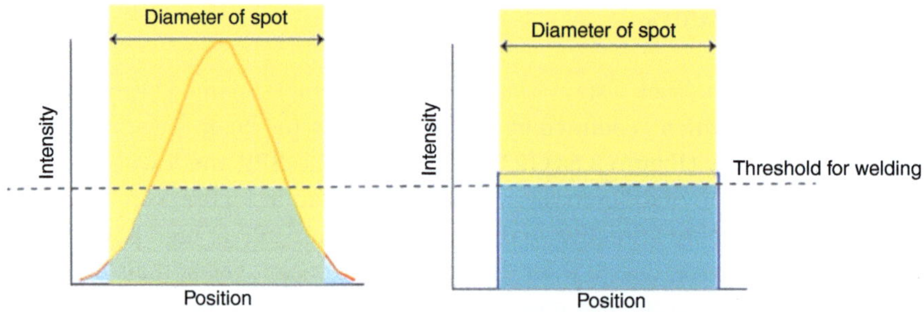

Figure 3.7 Profile of energy intensity distribution of (a) Gaussian and (b) top-hat laser beams. The blue areas show the portion of the energy used for welding.

In addition to the reflection of the beam, the energy of laser is wasted in another way. The plasma and the vapor may absorb or scatter the laser beam and hence lower the laser energy. When the plasma is not suppressed enough, the welding mode may change from deep penetration into heat conduction mode [26]. The characteristics of the plasma are dependent on the wavelength of the laser. The absorption of the laser beam in the vapor plume is proportionally dependent on the square of the laser wavelength. In CO_2 laser welding with a wavelength of 10.6 μm, the absorption coefficient is 100 times higher than that of the fiber laser with a wavelength of 1.07 μm. So, the vapor plume gets ionized in CO_2 laser welding, and hence, it has a high tendency to form plasma, while in fiber and Nd : YAG the vapor plume is less ionized [27]. Due to the absence of plasma in these processes, the scattering of the laser beam is low and the beam intensity distribution is narrow. In this case, the stability of the welding process is enhanced and the threshold power needed to induce the keyhole is lower [10]. However, plasma shielding effect in fiber laser welding needs to be considered when the laser power is greater than 5 kW, as the plasma plume varies from a weakly ionized metal vapor to a strongly ionized plasma beyond this power [28].

3.3 Defects Formation During Laser Welding of Al Alloys

3.3.1 Mechanism of Pore Formation

In addition to the pores that are made by metallurgical factors, another type of pores is made by the process named keyhole-induced porosity, which is due to unstable keyhole during LBW [29]. These pores are composed of metal vapor, entrained shielding gas, and air [29]. The laser beam melts the material locally and vaporizes it, causing a very deep and narrow hole to be created, which is called keyhole or vapor capillary. The pressure of the vapor prevents the closure of the vapor capillary. The walls of the capillary reflect the laser beam several times, causing more effective absorption of the laser beam by the melt. During laser welding with CO_2, the energy of the laser is absorbed by the vapor, causing it to be ionized, which forms the laser-induced plasma. This forms a plume plasma that can interact with the laser beam, leading to variation in the penetration [30]. A successful weld by

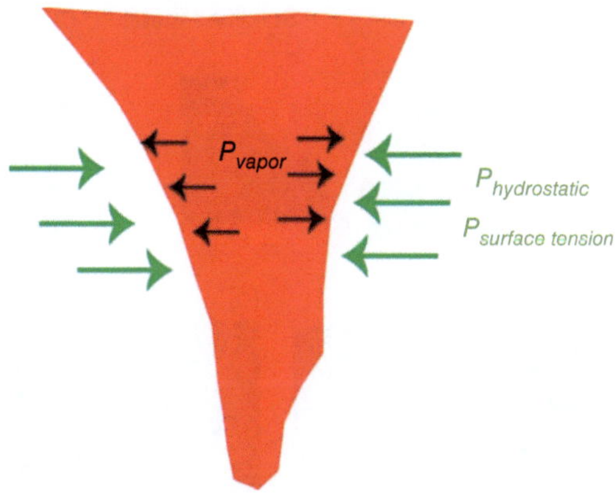

Figure 3.8 Schematic of pressures acting on the keyhole.

LBW requires a stable keyhole. The stability of the keyhole is obtained by a balance between the vapor pressure and the keyhole pressure. While the former tries to open the keyhole, the second closes it mainly by the mechanism of surface tension [31].

The mechanism of pores formation originating from the keyhole is explained by the pressure inside the keyhole [32]. The vapor pressure inside a keyhole at the bottom of it (P_v), also referred to as the recoil pressure (P_r), is in equilibrium with other pressures [32]:

$$P_v = P_a + \rho gh + 2\gamma/R \tag{3.3}$$

where P_a is atmospheric pressure, ρ is density of the liquid metal, g is gravitational acceleration, h is keyhole depth, γ is surface tension, and R is the radius of the keyhole. The second and third terms on the right side of the above equation are referred to as the hydrostatic pressure and the surface tension pressure, respectively. The schematic of pressures acting on the keyhole is shown in Figure 3.8 [33, 34].

The mechanism of the keyhole pore is explained by the collapse of the unstable keyhole [35]. This mechanism is explained in Figure 3.9. First, the pressure due to the surface tension increases, causing a projection to appear in the keyhole (Figure 3.9a). When this pressure exceeds the vapor pressure, the projection gets bigger (Figure 3.9b). Then, the gravitational force prevails on the liquid projection (Figure 3.9c, d) and finally the keyhole collapses. In fact, when the bulge formed at the rear of the keyhole reaches the front wall of the keyhole, the keyhole collapses. This bulging may also result from the clockwise flow of the melt at the rear of the keyhole. For the stability of the keyhole, it is necessary for the recoil pressure to be higher than the pressure resulting from the surface tension (P_h), so the keyhole remains open, avoiding the formation of the pore [34].

As the height-to-width ratio of the keyhole increases, the surface tension increases, which leads to the collapse of the keyhole and the entrapment of bubbles [33]. Figure 3.10 shows how the control of height-to-width ratio in a hybrid laser-arc welding reduces the pores in the FZ. In Figure 3.10a, the arc current is low and the laser power is high, causing a high height-to-width ratio. According to Eq. (3.3), the surface tension is high and thus the keyhole collapses. When the height-to-width ratio is lowered by increasing the current of the arc, this surface tension decreases,

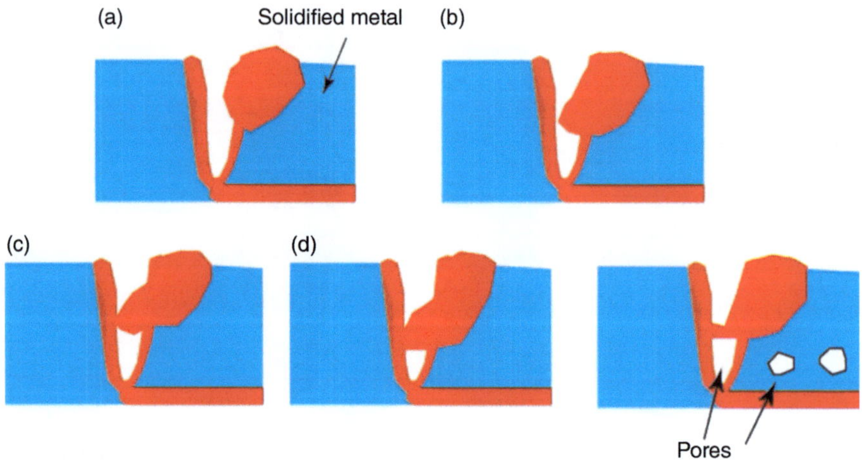

Figure 3.9 (a–d) Mechanism of keyhole-induced porosity. Source: Redrawn with permission after Ref. [35].

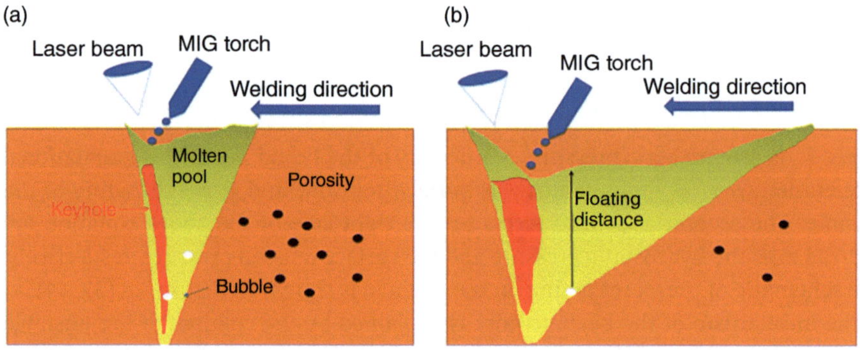

Figure 3.10 (a) High height-to-width ratio leads to higher number of pores in the FZ. (b) Lower height-to-width leads to lower number of pores. Source: Ref. [33].

leading to a formation of fewer pores (Figure 3.10b). This lower height-to-width ratio could be obtained by increasing the ratio of arc welding current to laser power in hybrid welding.

These bubbles contain gases whose composition depends on the mechanism of bubble formation. The bubbles at the bottom of the keyhole form due to extra evaporation of weld metal. These bubbles suffer expulsion due to the high pressure at the keyhole bottom and move to the back of the weld pool and are trapped at the solid–liquid interface (Figure 3.11a). This mechanism acts when the plume plasma is removed by using a fan, causing the keyhole to be stable. In this case, the porosities contain 100% hydrogen, which originate from the base material. The bubbles in the bottom of the keyhole contain metal vapor that finally moves to the interface of solid–liquid and is solidified. The vacuum formed in the pores causes the absorption of the hydrogen. When the plume plasma is not removed by a fan, the keyhole

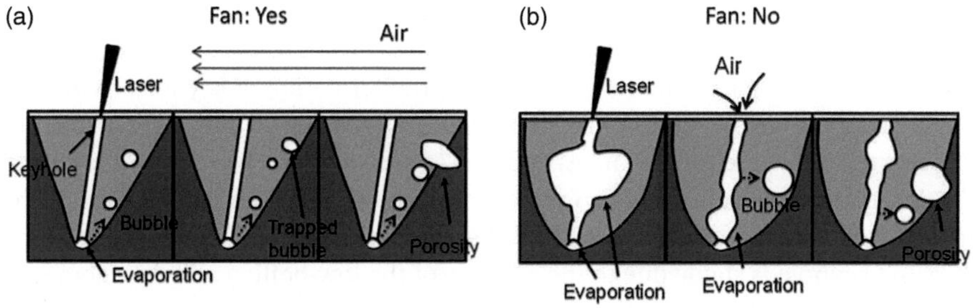

Figure 3.11 Mechanism of pore formation during laser welding of aluminum (a) with fan and a stable keyhole, (b) without fan and an unstable keyhole. Source: Ref. [32].

is unstable due to interaction of the beam and the plume. In this case, the air is penetrated into the keyhole, and therefore, the pores contain nitrogen in addition to hydrogen (Figure 3.11b) [32].

The bubbles that form in the FZ can be eliminated if they can float up and reach the surface before the completion of solidification. The relationship between the escape velocity (v) and the minimum radius of bubbles (R) is [20]:

$$v = \frac{2gR^2(\rho - \rho_0)}{9\eta} \tag{3.4}$$

where ρ and ρ_0 are the densities of the melt and the cavitation bubble gas, respectively. η is the coefficient of viscosity of the melt and g is the gravitational constant. The higher the radius of the bubbles, the higher the escape velocity, i.e. the larger the distance that the bubble can float up. Therefore, the smaller bubbles are more likely to be trapped. The critical radius of the bubbles below which they are entrapped is dependent on the solidification time as well as the type of the gas in the bubble, according to the above equation.

Some alloying elements such as zinc and magnesium found in some Al alloys such as 7xxx have a higher vapor pressure with respect to Al. This leads to loss of these elements during laser welding, which affects the strength and corrosion resistance of the weld zone. This also causes a phenomenon called spiking effect during laser welding through which the vapor pressure drops and the vapor capillary becomes unstable. This behavior in Al—Zn—Mg—Cu alloys results in a temporary vapor pressure drop and collapsing of the keyhole [31]. These alloys also exhibit low surface tension and low viscosity due to the presence of Zn and Mg, which aggravate the stability of the keyhole [31]. This spiking effect occurs at the root of the weld and in the front side, which leads to undercut formation at the top [31]. At the upper part of the weld, the pore formation is less likely to occur during laser welding of Al alloys containing magnesium. This is due to the high recoil pressure resulting from the evaporation of the magnesium [34]. In other words, in the weld zone of Al alloys such as AA5083, the pores are more concentrated at the bottom, while in Al alloys with lower content of magnesium, the pores are distributed more uniformly in the upper and lower parts of the weld [34].

Besides Mg, other alloying elements with low boiling temperatures lower than Al, such as Si, Mn, and Zn, affect the dynamic of the keyhole. The total content of these elements in Al alloys exceeding 2.1% causes a violent fluctuation in the keyhole due to unbalance between the recoil pressure and the surface tension. These elements evaporate earlier than Al and hence accelerate the formation of keyhole, its growth rate, the depth of the keyhole, and its diameter [14]. Among Al alloys, the porosities are observed more frequently in AA1xxx, AA5xxx, and AA7xxx. In AA5xxx and AA7xxx alloys, the main reason for pore formation during laser welding is the unstable keyhole due to vaporization of the alloying elements. Though in AA1xxx alloys the keyhole is stable due to the absence of the low-boiling alloying elements, porosities are found in the weld zone, which are attributed to the slow escape of the bubbles from the molten pool. So, the factors other than low-boiling alloying elements need to be considered for the prediction of porosity formation such as the laser parameters that affect the bubble escape [14]. When using filler wires for laser welding, to avoid the porosity, the filler wires with lower content of Mg have to be used. This selection becomes even more significant, considering that the molten filler wire itself disturbs the keyhole [36]. The mechanism of pore formation in five Al alloys and the total percentage of low-boiling elements in each alloy are shown in Figure 3.12a and b, respectively.

Al alloys are more prone to keyhole instability than steel during the laser welding. Al melt has a lower surface tension and viscosity than steel, and hence, the bulge formed at the rear of the keyhole reaches more easily to the front wall (Figure 3.9) and thus the keyhole collapses more easily [37]. Furthermore, the height-to-width ratio of the keyhole in Al is higher than steel due to lower melting temperature and higher thermal conductivity. Hence, the distance of the bulge at the rear from the front wall is smaller, making the collapse easier [37]. To suppress the formation of the porosities, the best way is to reduce the generation of the bubbles through a stable keyhole and reduce collapse of the keyhole or enhance the escape of the bubbles [38].

The size of the keyhole is also important in degassing of the gases during LBW. A narrow size of the keyhole causes the keyhole to collapse due to the prevalence of the surface tension and closing the hole. In this case, the pores are entrapped in the melt pool [31]. The thick Al sheets are more likely to develop keyhole instabilities due to the high hydrostatic pressure of the melt [31]. Preheating is

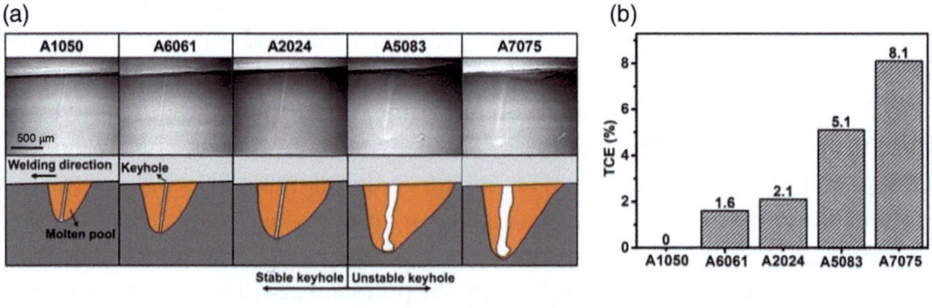

Figure 3.12 (a) Schematic presentation of the keyhole during laser welding of various aluminum alloys. (b) Total percentage of low-boiling-temperature elements (TCE%) in each alloy. Source: Ref. [14]/Springer Nature/CC BY 4.0.

shown to reduce the pores during laser welding of Al [39]. This is due to the lower surface tension of Al at higher temperatures, which makes it less likely to overcome the recoil pressure, and therefore the keyhole is more stable. The surface tension temperature coefficient for Al is negative according to the following equation:

$$\gamma = \gamma_0 + \frac{d\gamma}{dT}(T - T_0) \tag{3.5}$$

where γ_0 is the surface tension at the melting point temperature (T_0). The keyhole instability can also produce spatters. The spatters can be reduced when there is a stable keyhole, but they cannot be overcome completely, as the detachment of material from the wall of the keyhole by the vapor is inevitable [40].

3.3.2 Hot Cracking in Laser Welding

The susceptibility to SC in laser welding is more than arc welding. This is due to the high cooling rate, which from one side causes a high strain rate and stress gradient and from the other side causes the formation of a fine dendrite structure, which reduces the permeability [41]. The schematic of the solidification during laser welding of Al is shown in Figure 3.13. The middle of the weld between the initiation of the coherent zone ($f_s = f_{coh}$) and the end of solidification ($f_s = 1$) is prone to SC because the back-flow of liquid is more probably impeded and thus the volume change caused by shrinkage and thermal deformation cannot be compensated. When the strain rate in this region exceeds a critical value, the crack initiates. This critical strain rate is dependent on a parameter named depth-specific line energy (E_{Depth}), which is defined as [42]:

$$E_{Depth} = \frac{P}{vs} \tag{3.6}$$

where P, v, and s are the power of the laser beam, feeding rate, and the welding depth, respectively. The larger this value, the higher the critical strain rate and therefore the lower the SC susceptibility. At higher values of E_{Depth}, the number of grain

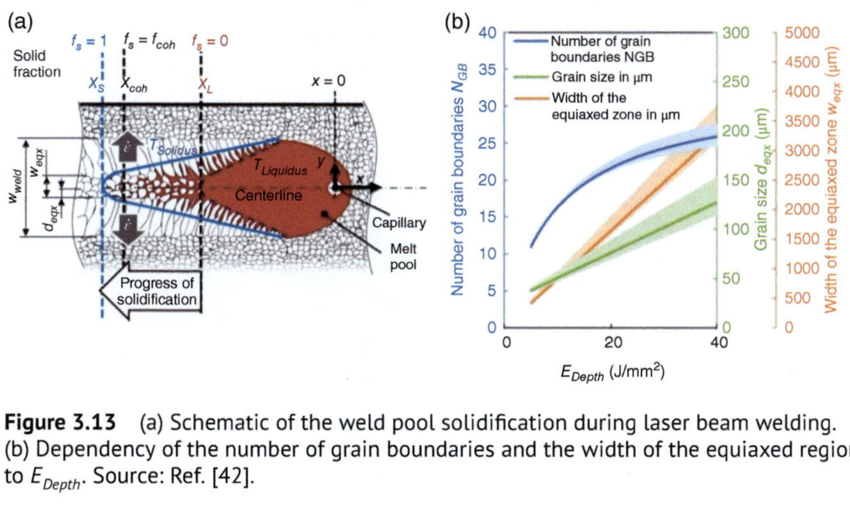

Figure 3.13 (a) Schematic of the weld pool solidification during laser beam welding. (b) Dependency of the number of grain boundaries and the width of the equiaxed region to E_{Depth}. Source: Ref. [42].

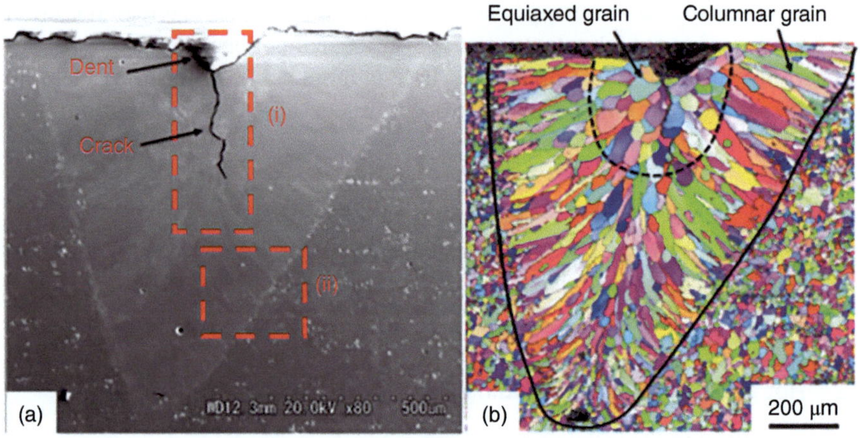

Figure 3.14 (a) Macrograph of a laser spot weld of AA5083 taken by SEM. The dent is due to solidification shrinkage. (b) EBSD of the weld zone showing the propagation of SC at the top surface with equiaxed grains near the dent. Source: Ref. [43]/Optica Publishing Group.

boundaries and the width of the equiaxed region increases, as seen in Figure 3.13b. In fact, with the increase in the width of the region with equiaxed grains (W_{eqx}) and the number of grain boundaries (lower d_{eqx}), the strain will be distributed over a wider area helping to inhibit the crack formation [42].

During laser spot welding, SC initiates and propagates on the upper surface of the weld at the final stage of solidification where a high solidification shrinkage exists. This shrinkage is characterized by a dent on the surface of the weld [43]. Though equiaxed grains in the middle area are less prone to SC, the SC propagated through this region due to the high strain and probably the strain rate. The center of the weld is the most critical region susceptible to SC (Figure 3.14). In Al alloys when welded in a lap configuration by laser welding, the SC starts in the root of the weld between the two sheets where the interface acts as a natural pre-crack (Figure 3.15a) [44]. A coarse columnar grain structure exists in this region which promotes the segregation of inter-dendritic low-melting compounds, promoting the crack growth parallel to the columnar grains [45]. This crack, when it reaches the centerline of the weld,

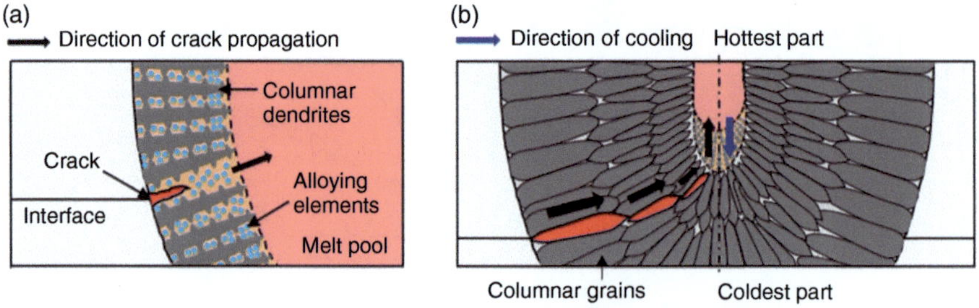

Figure 3.15 Schematic of SC during laser welding of Al alloys joined in a lap configuration. (a) Location of crack initiation from the roof of the weld at the interface. (b) Crack propagation to the center of the weld. Source: Ref. [45].

propagates upward like what is seen in butt joining, as the maximum strain and segregation of low-melting compounds favors this crack direction (Figure 3.15b).

When the grains are fine (at high *GR*) and equiaxed (at low *G*/*R*) and abundant eutectic liquid is available, the liquid between the grains can deform easily, lowering the susceptibility to SC [46]. Further grain refinement below a critical value by equiaxed grains increases the susceptibility to SC due to lower permeability [47]. Various strategies have been used to overcome this issue during laser welding of Al alloys. The control of grain size is a practical approach to reduce SC during laser welding of Al alloys. Grain refinement can be obtained either by controlling the solidification parameters (*G* and *R*) and chemical composition, or by adding the grain refiners. A filler material of AlTi5B1 during laser welding of AA6082 controls the grain size and fraction of equiaxed grains and therefore the susceptibility to hot cracking (Figure 3.16) [47]. The lower the grain size (not lower than a critical value) and the higher the fraction of equiaxed grains, the lower the susceptibility to SC.

The high welding speed promotes the SC during laser welding of Al alloys [48]. The critical speed above which the hot cracking occurs depends on various parameters, one of the most important ones being the chemical composition. When laser welding of dissimilar Al alloys, the melt ratio of two alloys plays an important role in the control of SC, as this ratio determines the chemical composition of the weld. To control this, the selection of the optimum laser welding parameters is critical [49]. Figure 3.17 shows how the critical welding speed for SC increases as the overlap width (*D* in Figure 3.17a) decreases during laser welding of AA6013 (upper sheet)/AA5754 (lower sheet) in a lap configuration. When *D* decreases, the fusion ratio (which is defined according to the following formula) increases and subsequently the contents of Mg and Si in the melt change, making the chemical composition less susceptible to SC. Therefore, the critical welding speed for SC increases (Figure 3.17c) [49]. The fusion ratio (*R*) is defined as:

$$R = \frac{S_L}{S_U + S_L} \tag{3.7}$$

where S_L and S_U are the fusion areas of the lower and upper sheets, respectively (Figure 3.17b).

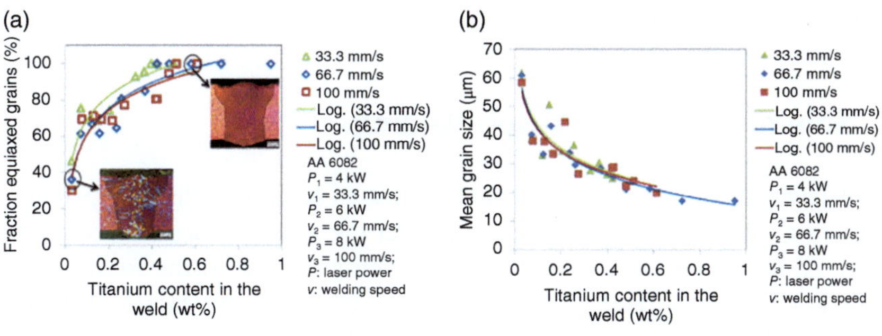

Figure 3.16 Effect of titanium content on the fraction (a) and size (b) of the equiaxed grains during laser welding of AA6082 at various welding speeds. Source: Ref. [47]/Springer Nature.

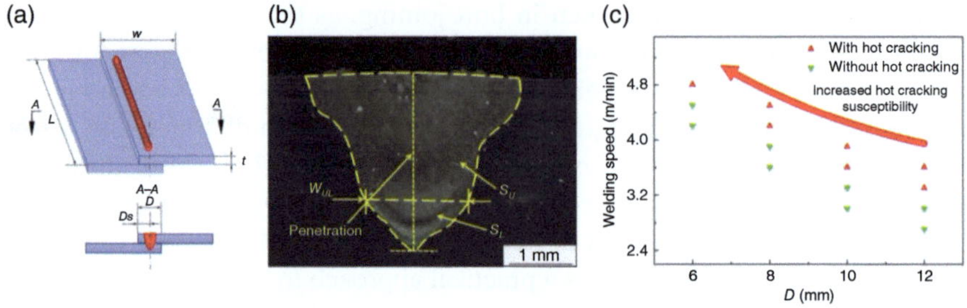

Figure 3.17 (a) Schematic of the laser welding of two aluminum alloys in overlap weld. (b) Weld cross section. (c) Dependency of the critical welding speed for the SC to the overlap width (*D*). Source: Ref. [49]/Springer Nature.

Another important factor during laser welding of dissimilar Al alloys in a lap configuration is the stacking order of two sheets. Depending on the position of the sheets (top or low) and the melt ratio, various chemical compositions can be obtained, which may or may not lie in the SC susceptibility range. When the less-sensitive alloy is placed on top, the fusion ratio should be lower than a critical value to avoid SC and vice versa [50].

Other strategies may help to reduce SC susceptibility. Applying a pressure in the transverse direction during laser welding can reduce hot cracking due to inhibition of the liquid film rupture [51].

Addition of Ti into the weld metal is another method to mitigate the SC [52]. The increase in the width of the equiaxed zone in the middle of the weld is the main reason that hinders the hot cracking. Ti in molten Al is present as solid particles of $TiAl_3$, which reacts with molten Al according to a peritectic reaction [53]:

$$\mathrm{Al\,(L) + TiAl_3\,(S) \rightarrow \alpha\text{-}Al\,(S)}$$

This reaction occurs in a way that aluminide particles are covered by solid α-Al that act effectively as nucleation sites for equiaxed grains.

Another important factor during laser welding of Al alloys that affects SC is the loss of alloying elements due to vaporization of volatile elements such as Mg and Zn [54]. Through vaporization, the chemical composition of the weld may fall in the susceptible composition range for hot cracking.

3.4 Shielding Gases in Laser Welding

The protective gas can affect the weld quality through several mechanisms. Helium or argon or a mixture of them is used to protect the weld pool against oxygen or nitrogen. Al has a high affinity to oxygen, which gives rise to the formation of an oxide layer on the surface of the molten pool, inhibiting the escape of the gas bubbles [55]. When using the protective gas, the sole source of hydrogen would be the oxide layer that previously existed on the surface of Al [27]. A high increase in the mechanical properties of Al welds was observed when laser welding was performed with a protective gas [55]. The mixture of He + Ar with the helium content above 50% yields a lower porosity in the weld zone by broadening the weld fusion and penetration [56]; especially in CO_2 laser welding, higher content of He is recommended as Ar absorbs the incident laser energy and defocuses the laser beam due to the large size of the laser plume [57]. N_2 as a protecting gas in laser-arc hybrid welding (LAHW) produces nitride flakes in the FZ, causing an anisotropy in the mechanical properties as well as a decline of the UTS and elongation [58]. These flakes cause the fracture pattern of the weld to consist of both brittle and ductile characteristics. On the other hand, LAHW under N_2 protective gas produces smaller grain sizes due to frequent short circuit of the arc weld and decrease in the temperature, which subsequently leads to higher yield strength (YS) of the FZ [58].

In addition, the shielding gas also undermines the plasma formation and the vapor over the weld bead [6]. The plasma over the weld interferes with vaporization of the metal, which in turn violates the keyhole formation [59]. The plasma causes a decrease in the laser absorption efficiency and unstable keyhole. The former causes a low penetration depth, and the latter leads to pore formation [59]. He has a higher ionization potential with respect to Ar, so the plasma effect is reduced when using He. In addition, higher heat and thermal conductivity of He produces a hotter weld pool with a longer solidification time, allowing the gas bubbles to escape and reduce the risk of gas entrapment [60]. When using a mixture of Ar—He in LAHW, the higher content of He causes a higher penetration and a lower porosity formation [59]. The higher penetration depth while using He is attributed to higher thermal conductivity of He.

In the absence of keyhole when the heat is transferred through conductivity, for instance in weld-brazing of dissimilar joining, the shielding gas can affect the wettability of the molten material on the surface of the other material. Addition of CO_2 to argon increases the wettability of Al over the surface of steel due to the formation of an oxide layer over the molten pool, which provides a pressure for Marangoni effect. This Marangoni effect causes a downward flow of the molten pool to the solid surface of steel, which improves the wettability [61].

Shielding gases may be entrapped and form pores in the weld zone [62]. The solubility of the protective gas influences the pore formation. For instance, when nitrogen is used for laser welding instead of argon, less pore is observed, which is attributed to the quick dissolve of nitrogen in the liquid Al in the weld pool [39]. The protective gases also promote SC due to their cooling effect on the surface, which results in a high thermal gradient [51].

Laser welding in a vacuum in comparison with laser welding in the atmosphere leads to smaller weld pool and higher viscosity due to the lower evaporation temperature. This causes a higher stability of the vapor capillary and a lower spatter. Furthermore, the vacuum hinders the formation of new oxide scale on the weld bead. The plasma plume is also suppressed by performing laser welding in low ambient pressure, which subsequently lowers the interaction of the laser beam and the plume [63]. So, by laser welding in the vacuum, lower energy input is required with respect to laser welding in the atmosphere with the same weld deep. The vacuum causes the boiling point of the melt (T_b) to decrease according to Clausius–Clapeyron equation [64]:

$$\frac{d \ln p_{ambient}}{dT_b} = \frac{\Delta H_{vap}}{RT^2} \tag{3.8}$$

where $p_{ambient}$ is the ambient pressure, H_{vap} is the latent heat of evaporation, and R is the gas constant. While there is no change in the melting point, a high drop in the boiling point of the melt means that the temperature difference between the keyhole wall (whose temperature is about the boiling point) and the weld pool boundary decreases with decreasing ambient pressure [64]. This reduces the elongation of the weld pool in the welding direction (Figure 3.24). In other words, the molten pool around the keyhole is thinner. This along with the lower temperature of the melt causes it to have a higher surface tension and viscosity, which leads to higher stability of the keyhole [65]. As the keyhole wall can be formed at a lower temperature, less energy is needed to form the keyhole. Besides, the evaporation rate from the keyhole wall would be higher, which means a deeper keyhole and a higher recoil pressure are produced. This also affects the melt flow path when welding in a vacuum, which helps the pores to escape more easily when welding is performed in a vacuum [64].

3.5 Modifications to Laser Welding of Al

3.5.1 Multi-spot Laser Beam Welding

There is a high demand on the appearance of the car body parts that are visible. The weld seams in these areas need to be smooth so they can be painted without any reworking. Spot-in-spot configuration is a good choice for this purpose, in which a main laser spot with a high power is surrounded by another spot, which smoothens the weld bead in the same operation. By multi-spot module, it is possible to make various spot geometry or split the laser beam into several spots with adjustable distances and varying intensities within each spot. By spot-in-spot configuration, both symmetric and asymmetric weld beads can be obtained with higher welding speed and better welding quality with respect to conventional circular or rectangle spots. Using several spots helps to form the shape of the vapor capillary in order to reduce the porosity [17]. Concentrically arranged spots with various powers can be used to control the porosity [17]. The change of the power distribution in the core and in the outer ring can effectively reduce the pores.

Muti-focus laser beam causes an increase in the keyhole area and an enlarged weld metal pool with a more stabilized keyhole, which prevents it from collapsing [3]. Superposition of one laser system on another laser system is a useful procedure to alleviate SC during laser welding of Al alloys. Figure 3.18 shows the schematic of the superimposed diode laser on pulsed Nd : YAG laser welding. Figure 3.18a–c shows the cross-sectional images of this weld performed on a 6xxx alloy at various powers of the superimposed diode laser. The decrease in the SC with the increase in the power of the diode laser is due to the lower cooling rate and longer solidification time of the weld pool [66]. An increase in the weld width and depth due to higher heat input is also clear.

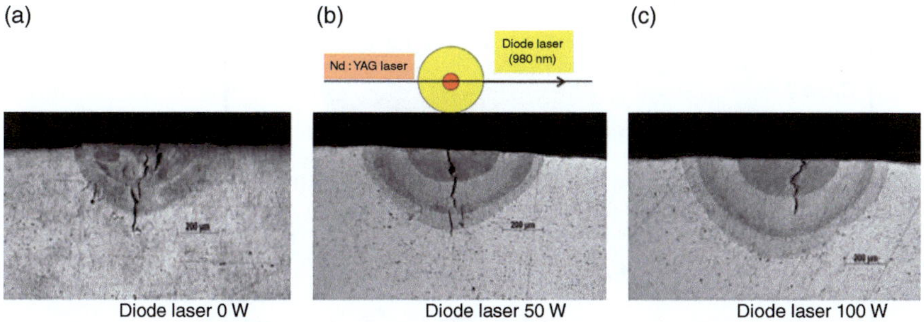

Figure 3.18 Schematic of the superimposed diode laser on pulsed Nd : YAG laser welding. (a–c) The effect of various diode laser powers on the hot crack susceptibility. Source: Ref. [66]/with permission of Elsevier.

Multi-focus beam shaping is a technique used to distribute energy input in the keyhole in different ways. By setting several focal positions inside the keyhole, the energy is distributed more uniformly in the entire depth of the keyhole. In this way, the middle of the keyhole is less collapsed and less protrusions are produced. So, the obstacles in the way of the vapor are reduced, causing lower detachment of the wall of the keyhole and therefore lower spatter. Additionally, by setting one focal in the upper part of the keyhole, the opening becomes big enough to allow the vapor to escape more easily, causing even less spatter [67].

3.5.2 Micro-laser Welding

This process can join the sheets within the thickness range of 0.03–1 mm. The duration of pulses is in the range of 1–50 ms and the frequency is 1–50 Hz in the way that the weld pool solidifies during the pause of the pulse [41]. This process is used to join Al sheets in the microelectronic and energy industries [68]. The infrared fiber laser sources allow to attain the spot sizes below 30 μm, which is very useful in the battery tab welding, because the chemicals inside the battery are very sensitive to heat [69]. Due to the small diameter of the laser spot, a very narrow weld is formed, which can be resolved by using spatial power modulation or wobble technique. In this technique, a linear feed rate is superimposed with a circular oscillating motion (Figure 3.19).

3.5.3 Pulsed Laser Beam Welding

In conventional laser welding, a continuous laser beam with a defined constant power during the welding process is applied, while in pulsed LBW, the laser beam is emitted in sequentially switched pulses [41]. With a single pulse, a spot welding is realized. When the spot welds are overlapped, a seam weld is obtained. An overlap of more than 70% is required to achieve the desirable strength. The pulse overlap is obtained through the following equation [68]:

$$Overlap = 1 - \frac{Welding\ speed \cdot Pulse\ period}{Diameter\ of\ the\ spot\ weld} \tag{3.9}$$

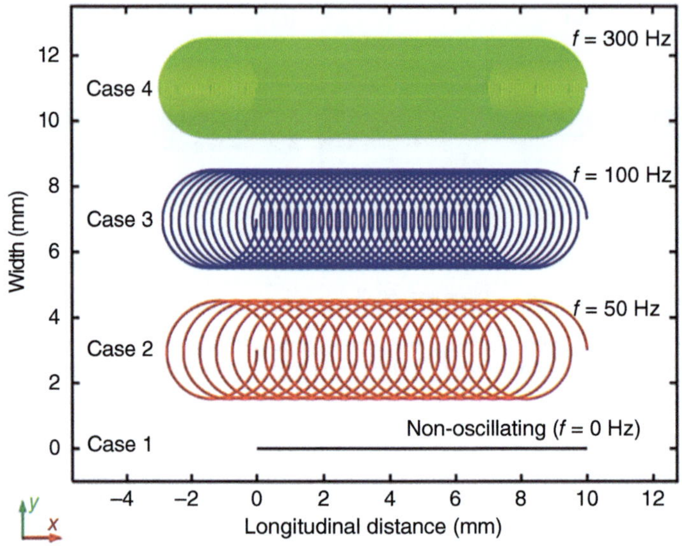

Figure 3.19 Schematic of the wobble technique showing various frequencies. Source: Ref. [70].

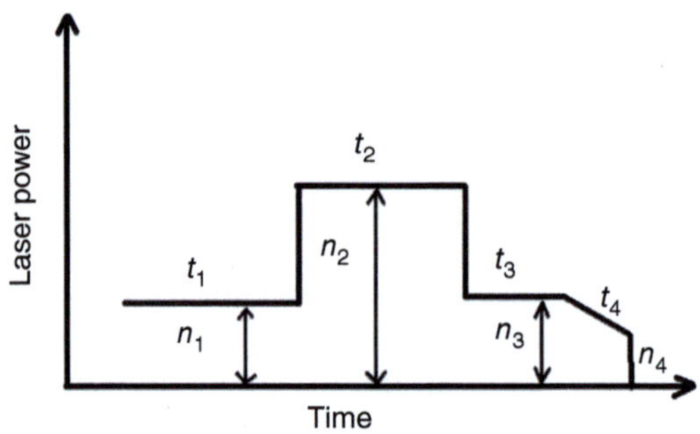

Figure 3.20 Schematic of the laser pulse shape.

Pulsed LBW is usually used for thin sheets as it can control the heat. The low temperature in the vicinity of the weld due to the high cooling rate of the pulsed beams helps to reduce the distortion in the thin sheets. Different pulse shaping with different power levels and time durations can be used to preheat the sample (n_1 in Figure 3.20). Other pulses can be used for welding (n_2) and, if necessary, for the cooling stage (n_3–n_4). This strategy can be used to control the aspect ratio of the weld pool to avoid SC [71].

In laser spot welding, the shape of the pulse can be optimized to control the SC [41]. When the shape of the pulse is triangle, SC is pronounced. With a ramp-down stage in a specific time of ramping, the hot cracking is avoided (Figure 3.21). When using a triangle pulse shape, the available time for solidification is very low and so the remaining melt cannot fill the cavity in the inter-dendritic space (backfilling).

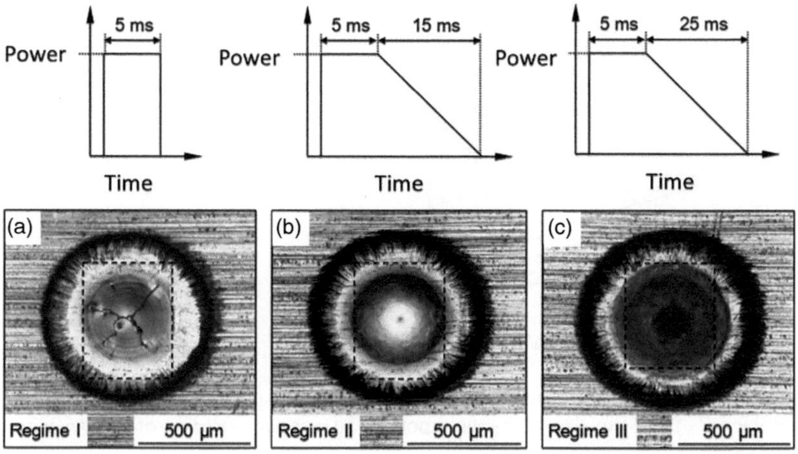

Figure 3.21 Various shapes of ramp-down stage in one pulse of laser spot welding. The images (a–c) show the corresponding surface of the spot welds.
Source: Ref. [41]/Ilmenau University Press.

In addition, the solidification region is enlarged due to lengthening of the dendrites or cells. These dendrites or cells are fine and have low permeability and thus further aggravate the backfilling process. Using a ramp-down stage in the pulse shape eliminates these effects and therefore the hot crack disappears (Figure 3.21b). However, increasing the time of the ramp-down intensifies the segregation, which promotes SC (Figure 3.21c) [41].

The slope of the ramp-down has another effect on SC due to its effect on the strain rate. The strain rate during welding is obtained via [21]:

$$\dot{\varepsilon} = -\beta_T \dot{T}_{SL} = \beta_T |G_{SL} V_{SL}| \tag{3.10}$$

where β_T is the thermal contraction coefficient due to shrinkage and thermal contraction and $\dot{T}_{SL}$ is the cooling rate at the interface of solid–liquid. $\dot{T}_{SL}$ is obtained by multiplication of temperature gradient at the solid–liquid interface (G_{SL}) and the velocity of the movement of this interface (V_{SL}). A decrease in the slope of the ramp-down at the final stages of solidification causes a decrease in the interface velocity (V_{SL}) and therefore a decrease in the strain rate, which subsequently reduces the SC susceptibility [21].

Another important factor in SC during laser spot welding of Al alloys is the size of the weld pool, which affects the strain in the weld pool (ε_{areal}) according to the following equation [21]:

$$\varepsilon_{areal} = \frac{A_{molten} - A_{solid}}{A_{solid}} \tag{3.11}$$

where A_{molten} and A_{solid} are the area of the spot weld at the beginning and after the solidification, respectively. At high power of laser welding where the welding mode changes from the heat conduction to keyhole, a sudden increase in the strain is observed, which promotes the hot cracking [21]. Figure 3.22 summarizes the effect of solid–liquid interface velocity and the size of the weld pool on the SC susceptibility.

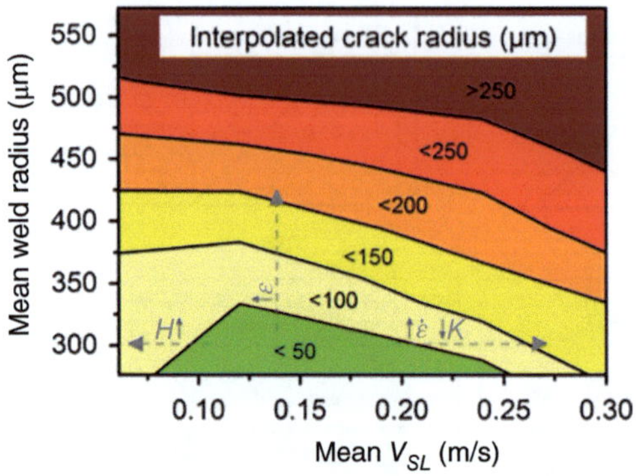

Figure 3.22 SC susceptibility (measure by crack radius) with respect to the solid–liquid interface velocity (V_{SL}) and the size of the weld pool in a laser spot welding of an AA6082 alloy. Source: Ref. [21].

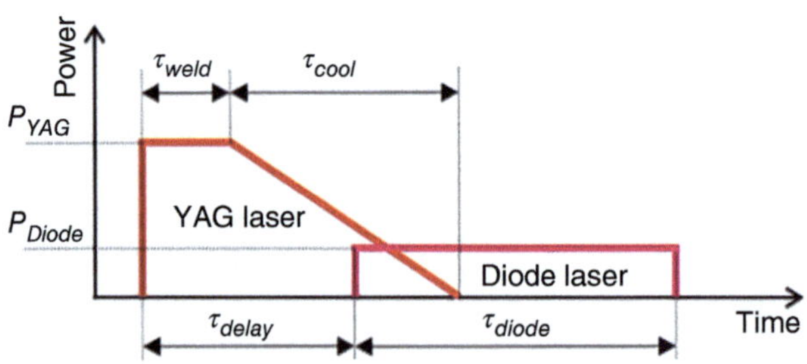

Figure 3.23 Shape of double-pulse laser welding. Source: Ref. [72].

In this diagram, SC susceptibility is quantified with the solidification crack length. By increasing the V_{SL}, the SC is promoted due to higher strain rate while by decreasing the weld radius SC is reduced due to lower strain.

Where the application of ramp-down of the laser power fails to overcome the SC, the superimpose of another laser beam is an alternative solution (Figure 3.23) [72]. In Figure 3.23 a low-power diode laser is applied at the final stage of the ramp-down of a YAG laser. The time delay of this pulse diode laser is crucial in alleviating the SC. The main mechanism behind this elimination of SC is the development of compressive stresses, which compensate the internal strains caused by solidification shrinkage and thermal conduction [72]. This second pulse does not affect the solid–liquid interface velocity, the grain size, and the dendrite microstructure, as it is applied right before the completion of solidification. Instead, it affects the thermal gradient and reduces the cooling rate, which causes a lower strain rate during the critical final stage of solidification.

The increase in the dendrite arm distances to facilitate permeability of the remaining melt during seam welding of Al can be obtained by lengthening the solidification time or, in the other words, by decreasing the solidification rate. During laser pulse welding by Nd : YAG, imposing another laser beam by diode laser with a lower power and a larger diameter causes the weld pool to be widened, which in turn reduces the

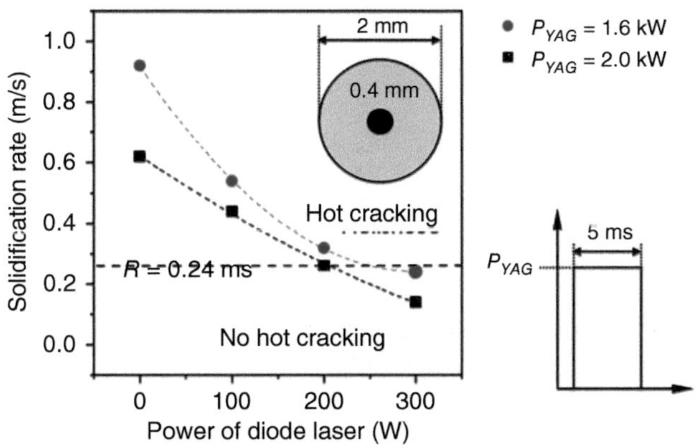

Figure 3.24 Effect of the power of the imposed laser beam as well as the power of the main laser beam on the solidification rate and SC susceptibility of an AA6082. Source: Ref. [41].

solidification rate [41]. Figure 3.24 shows how the increase in the imposed pulse reduces the solidification rate. The same effect can be obtained by increasing the power of the main laser beam (increase from 1.6 to 2.0 kW). The higher the power of the imposed laser, the lower the tendency for hot cracking. Below a critical solidification rate (0.24 m/s), no SC is observed.

Pulsation may have other effects in the weld pool. High-frequency modulation of the pulse beam enhances a convection Marangoni flow in the melt, which results in a constitutional undercooling in the melt boundary. Therefore, a fine equiaxed microstructure develops [73]. The lower heat intensity of pulsed laser beam reduces evaporation of some alloying elements such as Mg in the FZ, which can reduce its decremental effect on the hardness of the FZ with respect to conventional LBW [1].

3.5.4 Hybrid Laser Welding

Other welding processes may be used along with laser welding for different purposes. LAHW uses TIG or MIG beside the laser beam. The arc can be in the trailing or leading side (Figure 3.25).

One of the main reasons for using LAHW is the gap bridging, which can be up to 50% of the sheet thickness [57]. The tolerance of LAHW to misalignment, gap, and groove is lower than LBW. The arc cleaning effect helps to remove the oxide layer on the surface of Al, eliminating the disadvantages of LBW. The heat input of LAHW is lower than arc welding and higher than LBW. LAHW processes result in a more stable welding process with a deeper penetration and higher speed of welding [74]. Hybrid laser/MIG welding of Al is shown to produce less pores due to the higher heat input and the lower solidification rate, which facilitates the escape of the bubbles from the weld pool to the free surface [39]. Other newly developed arc welding processes such as CMT can reduce the detrimental effects of the heat of the MIG process. In this way, laser/CMT hybrid welding offers advantages over laser/MIG hybrid welding, such as finer microstructure and a narrower columnar dendrite zone [75].

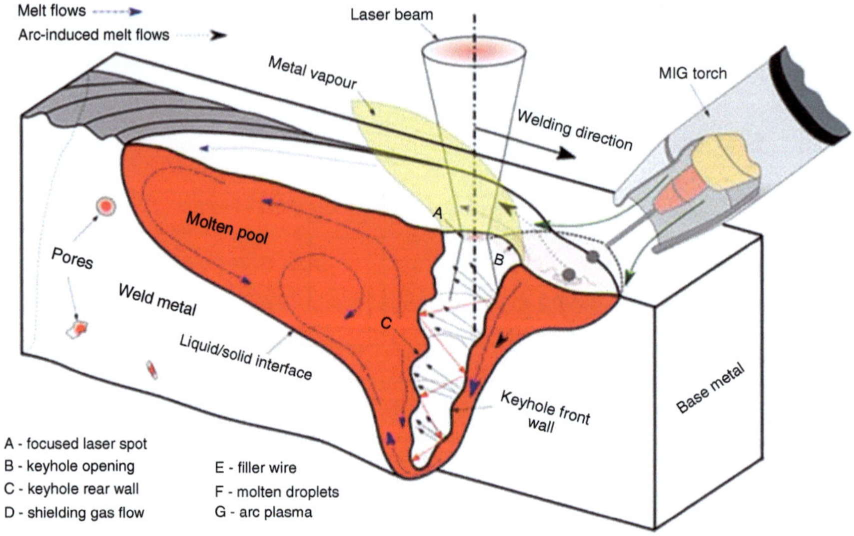

Figure 3.25 Schematic of LAHW with a leading arc. Source: Ref. [57].

Plasma welding is also used along with laser welding for thin sheets of Al alloy to increase the welding speed and, at the same time, to increase the penetration depth. A proper selection of the distance of the plasma arc with the center of the laser beam causes a high absorption of the laser beam by the surface of Al [76].

3.5.5 Oscillation Laser Beam Welding

Oscillation refers to the movement of the laser beam. The main parameters in this type of LBW are amplitude, frequency, and shape of the oscillation. Recently, high-power galvanometer scanners have promoted this technique. The microstructure of the weld zone can be refined and turned to equiaxed microstructure by proper design of oscillation. Oscillation of the laser beam can reduce the temperature gradient in the molten pool. This small temperature gradient favors the formation of equiaxed grain [77]. In addition to this, the stir of the melt flow penetrates into the mushy zone, causing the dendrites to beak by tiny turbulences. The broken grains act as nuclei for the growth of equiaxed grains (Figure 3.26).

The keyhole that is produced during welding is more stable by oscillation as the ratio of depth/width is lower [78]. Beam oscillation is reported to reduce the pore formation by various mechanisms. By oscillation, the rotation of the melt flow is steadier, which reduces the entrapment of shielding gas [78]. Continuous preheating and remelting of the FZ mitigate the pore formation [78].

High-frequency change of the position of the acting laser beam causes the keyhole opening to be larger, allowing the metallic vapor to escape, which reduces the keyhole-induced pores [2]. This oscillation can be a combination of distinct oscillations, such as a vertical one and a rotating one. Figure 3.27a and b shows how the keyhole-induced pores are reduced by applying a combined oscillation.

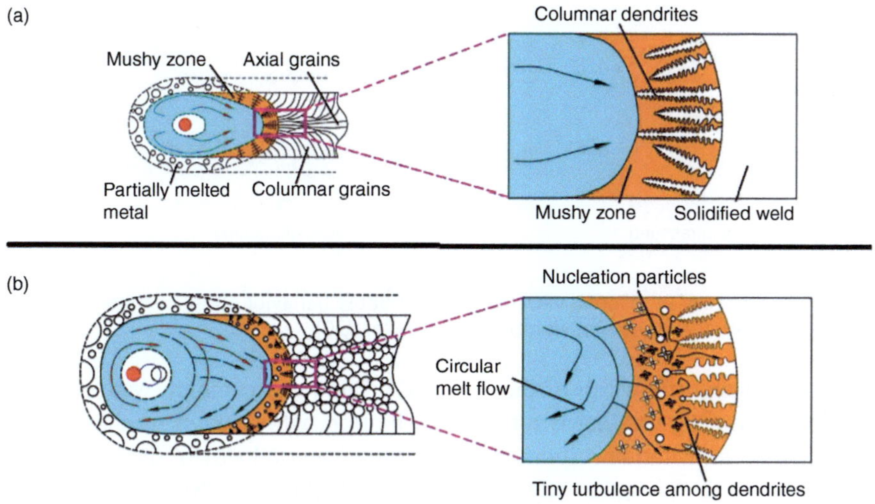

Figure 3.26 Schematic of the influence of the melt flow on the microstructure during (a) conventional laser welding and (b) oscillation laser welding with circular movement. Source: Ref. [77].

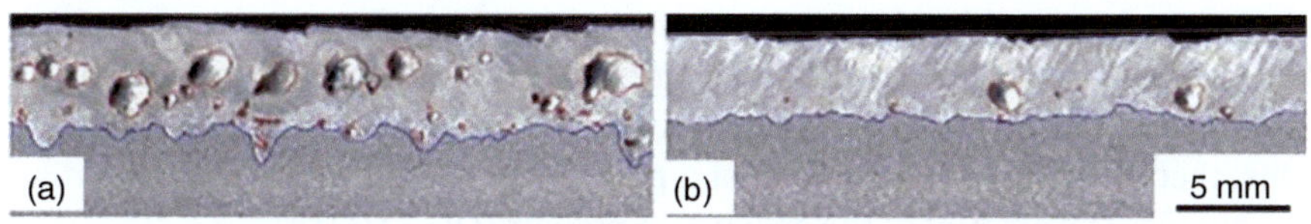

Figure 3.27 (a) Pores in the absence of oscillation. (b) Reduction of pores in the presence of combined oscillation. Source: Ref. [2]/MDPI/CC BY 4.0.

In this method, the rotation of the beam widens the keyhole, which helps the metallic vapor to easily escape the hole and reduce the keyhole-induced pores [79]. At the same time, it reduces the depth, which has a negative effect on the weld quality. A vertical oscillation component compensates this negative effect by increasing the weld depth [2].

Oscillation may also affect the stability of the keyhole by adverse melt flow behind the keyhole. The melt flow behind the keyhole in conventional LBW is upward, which causes bulging and bubble formation. The adverse downward flow behind the keyhole suppresses this bulging, but on the other hand it hinders the escape of the bubble upward. Overall, the probability of the porosity formation decreases with oscillation, as the main source of bubble formation (the instability of the keyhole by bulge formation) is mitigated [38].

3.5.6 Vibration-Assisted Laser Beam Welding

The vibration during LBW applied by ultrasonic can accelerate degassing [80]. Vibration increases the ratio of width/depth, which subsequently reduces the porosity [81]. Vibration can assist in modifying the microstructure of the weld zone as

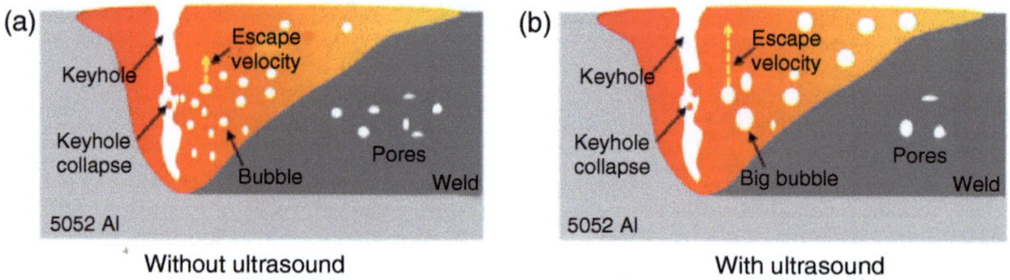

Figure 3.28 Effect of ultrasound vibration on the bubble behavior during welding: (a) without vibration and (b) with vibration. Source: Ref. [20].

well as decreasing the width of the HAZ [82]. The most common way to apply the vibration during welding is ultrasonic vibration, whose main mechanism in the weld pool is cavitation. Vibration causes the escape velocity of the bubbles to increase due to the cavitation effect of the vibration (Figure 3.28) [20]. According to Eq. (3.4), when the bubble radius increases, the escape velocity increases and thus more cavitation bubbles can escape before solidification and thus the porosity is reduced. Vibration causes a steady cavitation effect by which the bubble enlarges. By increasing the ultrasonic power, the number of broken bubbles increases and their size decreases, which decelerate the degassing process [83].

In addition to its effect on the pore formation, vibration assists in the modification of the weld microstructure and consequently the hot cracking. Around the cavitation bubble, there is a maximum pressure at a position of 1.587R, which R is the radius of the bubble. The amount of this maximum pressure ($P_{\max}$) is [84]:

$$P_{\max} = 0.15P_0\left(\frac{R_m}{R}\right)^3 \tag{3.12}$$

where R_m is the maximum radius of the bubble and P_0 is the static pressure. The shock pressure generated by the cavitation effect referred to as the acoustic stream has a stirring effect in the melt, which breaks the primary dendrites of Al and enhances the equiaxed microstructure [20]. The cavitation bubble also absorbs heat during their expansion, which leads to local undercooling, further promoting nucleation of the equiaxed grains [84]. The vibration is also effective in control of hot cracking. This goal is achieved through refined equiaxed grains which form in the centerline of the weld due to vibration [52].

3.5.7 Hot-Wire Laser Welding

The filler material in laser welding may be used to control the pore formation and chemical composition of the weld [85]. The filler wire consumes the laser energy to be melted, and hence, the full penetration of the process is not obtained [86]. To compensate for this, hot-wire laser welding is used. In this method, the electric current passes through the filler wire and heats it by electrical resistance, before it is melted by the heat of the laser. In this way, lower energy of the laser beam is needed to heat and melt the wire, making it more efficient. A stable melting of the wire needs a precise control of the tip temperature of the wire to be matched with the WFS and the laser power [87]. The heat input in this process becomes lower, the extent of the HAZ becomes smaller, the distortion diminishes, and higher welding speed can be selected to obtain a weld bead with a high quality. When this process is used for AM, the cyclic heating and remelting cause a coarsening in the microstructure adjacent to the melt pool boundary, which manifests itself as serration in the stress–strain curves [88]. These serrations are due to localized strain in these regions.

The battery packs in modern electric cars are placed inside the Al cases. All the components of the battery boxes need to be joined in the production process. The conventional welding processes such as MIG need a lot of time and cause huge amount of distortion, which is not tolerable in the assembly stage of the automobile. Hot-wire laser welding is a good solution for this purpose, which is also referred to as the "Precision Power Laser" from Lincoln company. Due to absence of the limitations of the arc, the welding speed is higher than MIG. Moreover, the weld bead characteristics are more controllable by this method. The proper combination of hot-wire and the laser makes this process usable for brazing, cladding, and AM.

3.6 Special Cases of Al Welding

3.6.1 Additive Manufacturing

Laser powder bed fusion (LPBF) facilitates production of complex geometries without limitations. Though the machines for AM have become larger in recent years, they are too small for the production of some large parts. Therefore, sometimes it is needed to join the parts made by AM with other existing parts. In these cases, combining AM with other conventional production methods can save time and cost. Conventional welding methods such as TIG or laser welding lead to pores formation during welding of additively manufactured parts [89]. Laser welding in a vacuum can overcome this problem. Physical effect of the reduced evaporation temperature and smaller weld pool causes lower pores from the base material to agglomerate and build large pores in the weld. The mechanism of pore reduction by various oscillation patterns of the laser beam is explained with regard to the heat energy and its influence on the bubble formation at the rear of the keyhole [90].

3.6.2 Cast Aluminum

The cast products of Al alloys, which are often produced by die casting, are usually in the form of semi-finished products. Die casting is a reliable method for the production of Al parts in series production in the automotive industry. These products contain entrapped gases or pores, which make them less weldable [91]. These pores contain hydrogen, oxygen, and nitrogen. The main sources of hydrogen are the mold release agent and organic piston lubricant. The high heat input of laser welding favors outgassing of hydrogen from the cast base material. Also, a low viscosity of the melt causes expulsion of the melting pool [92]. The usage of laser welding in a vacuum can overcome the pore building during welding of cast Al alloys, despite the presence of pores in the base material. Furthermore, a finer and more homogenized microstructure can be obtained in the weld zone, in contrast to the coarse eutectic structure in the base material. Application of an oscillating electromagnetic field produces a Lorentz force in the weld pool, which pushes the bubbles upward and reduces the porosity when laser welding a cast Al alloy [93].

3.6.3 Welding of Aluminum to Dissimilar Materials

Weld-brazing is a technique used to join Al to steel in a lap configuration. In this method, the heat of the laser beam melts the Al along with a filler material of Al, which wets the surface of steel and makes a metallurgical bonding [61]. For doing so, the laser should be defocused and the heat transfer is done by conduction without the formation of a keyhole [61]. In the weld-brazing process, the molten Al is in contact with the surface of the solid steel and the melting of steel should be avoided, as this accelerates the diffusion and interaction of Al and steel. This process can also be performed in a butt configuration using the hybrid arc laser process, as the arc helps to increase the wetting area. Figure 3.29a and b shows the cross sections of the joints made by weld-brazing between Al and steel in a butt configuration using laser and hybrid arc-laser techniques, respectively. No melting of steel is observed in these figures. The wetting of Al on steel surface is enhanced using an arc along with laser beam, which is due to higher heat input [94].

By increasing the interaction, the fast growth of IMCs will embrittle the joint [95]. In this state, the procedure is "welding," as the two metals are melted [96]. The cross sections of the lap joints between Al and copper made by laser welding in Al/Cu and Cu/Al arrangements are shown in Figure 3.30a and b, respectively. The melt-through of the bottom layer is visible in these figures, indicating the occurrence of welding. These SEM images clearly show that the arrangement of layers influences the type of IMCs as well as the location where they are present, which consequently affects the mechanical behavior of the joint [97]. The IMCs layer also affects the contact electrical resistance, though this property is less affected by IMCs than mechanical properties [98].

For weld-brazing of Al to steel, usually a coating of zinc on steel is used before the process, as the zinc layer facilitates the wetting of molten Al on the surface of

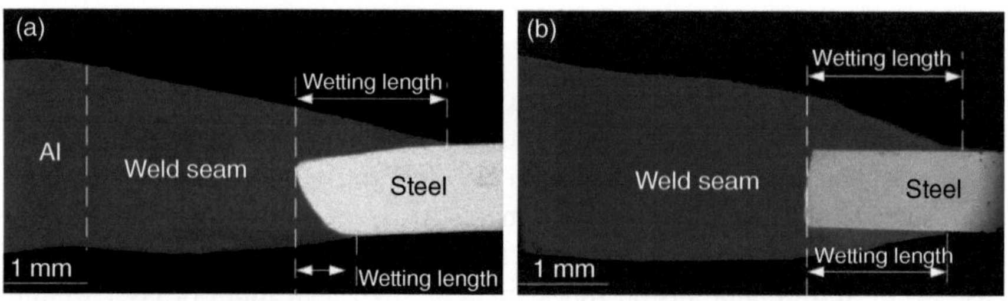

Figure 3.29 SEM images of cross section of weld-brazed joints between Al and steel in butt configuration made by (a) laser weld-brazing and (b) hybrid arc-laser weld-brazing. Source: Ref. [94]/MDPI/CC BY 4.0.

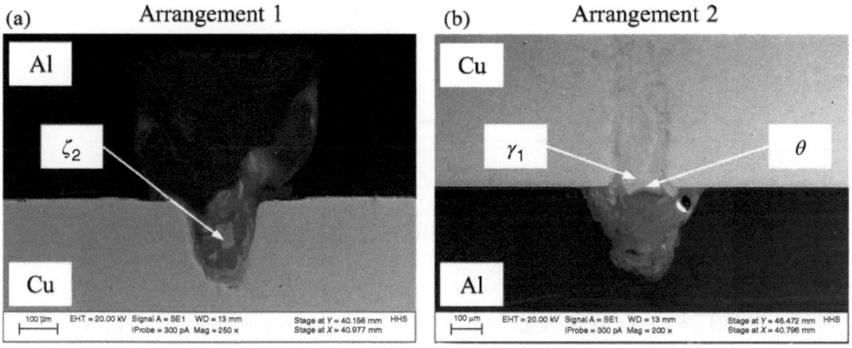

Figure 3.30 SEM images of cross section of Al/copper joints made by laser welding (a) when Al is on top and (b) when Cu is on top. Source: Ref. [97]/Springer Nature/CC BY 4.0.

steel [99]. The filler that is used for laser weld-brazing is zinc-based such as Zn-15%Al [100]. During this process, the heat of the laser beam is directed toward Al to avoid the vaporization of the zinc as well as the melting of the steel [100].

Al can also be joined to non-metallic materials such as polymers by another modification to the laser welding process. Laser transmission welding (LTW) is used to join two materials, mainly in a lap configuration by fusion at the interface. This requires the upper material to be transparent to the laser beam. In this way, the laser radiation can pass and heat the lower material. This process is used to join polymers to polymers as well as to join polymers to metals and hence is referred to as laser plastic welding [101]. It is also used to join glasses to metals for sealing purposes. An absorber may be used at the interface to facilitate the conversion of light energy into thermal energy [102]. In the case of glass/Al joining, a micro-arc-oxidation coating increases the laser energy absorption [103]. Due to different intrinsic structures of glass and metal, the mechanical bonding is the dominant mechanism of bonding, with some evidences of chemical reaction between Al—Si—O [103]. In polymer/Al joining, the formation of bubbles affects the joint quality [104]. The absorption of laser beam for natural polymers is low in the wavelengths range from 0.35 to 1.4 μm, making Nd:YAG, diode, and fiber laser good choices for LTW [105].

3.7 Simulation of Laser Welding

Simulation of the LBW is performed for several purposes. Like arc welding processes, thermal-mechanical simulation is used to calculating the residual stress and distortion. Residual stresses affect the mechanical performance of the joints [106]. Despite the lower heat input of laser welding and usage of clamps, longitudinal and transverse distortions appear during laser welding of Tee-joints. The schematic of longitudinal and transverse distortions is shown in Figure 3.31a and b, respectively. The out-of-plane deformation may occur during welding in a lap configuration [107]. During butt laser welding of thin sheets, weld bowing or angular deflection

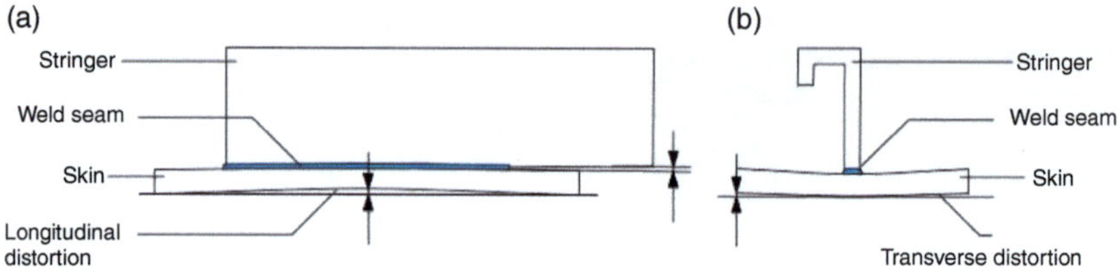

Figure 3.31 Schematic of distortion made during laser welding of Tee-joints. Source: Ref. [6].

may occur [108]. An accurate calculation by simulation to capture more detail requires consideration of metallurgical models to take into account softening and phase transformation (when welding to steel) [107]. Compared with experimental methods, simulation is a cheaper method and a strong tool to calculate residual deformation as well as residual stress of welding. Developments in machine learning techniques and using these approaches to model the simulation results made it even faster and less costly to optimize a lot of goals in the welding process with a lot of parameters [109].

Simulation methods yield some results that are not obtained by experiments. They also help to interpret the mechanism behind microstructural changes and defects formation. For example, the heat flux vector and the distribution of residual stress in various parts of the weld zone obtained by simulation help to interpret the grain orientations in the weld zone and hot crack initiation, respectively [110]. Thermal-mechanical FE analysis is a common approach to calculate the residual stress during laser welding. Cylindrical plus a Gaussian heat source is an example used to model the heat source in the keyhole (Figure 3.32). The parameters of the heat source that is used for laser welding need to be calibrated through matching the weld cross sections of the experiment and simulation [111].

Contour method is another method to measure the residual stress during laser welding of Al alloys. The fundamentals of this procedure are described in Chapter 2. The curves of distribution of longitudinal residual stress (normal to the welding plane and parallel to the welding direction) obtained by both contour and

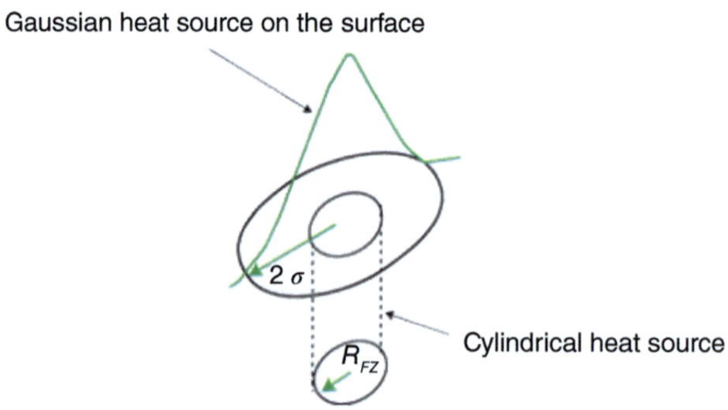

Figure 3.32 Schematic of a cylindrical plus a Gaussian heat source used to model the keyhole in laser welding. Source: Ref. [111].

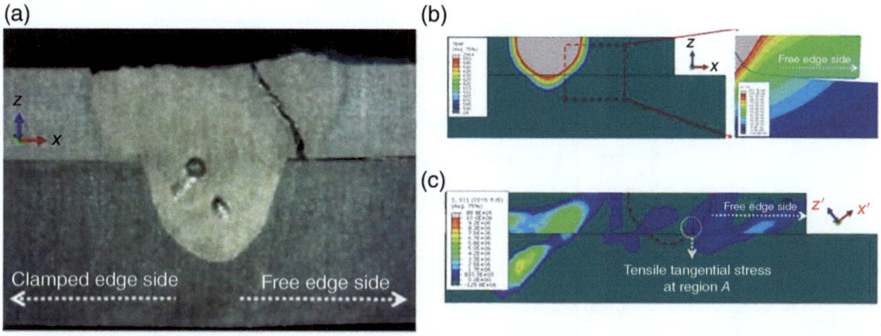

Figure 3.33 (a) Hot cracking observed during laser beam welding of two aluminum sheets in a lap configuration. The free edge is on the right side. (b) Mushy zone region where the temperature is in the range of 590–650 °C along with the interfacial deformation. (c) Stress distribution in the direction of *x*′. Source: Ref. [113]/AIP Publishing.

thermal-mechanical FE methods show a good match between the two approaches. A high residual tensile stress is seen in the middle of the weld, decreasing as the distance from the weld center increases [112].

Simulation of the laser welding can also help to understand the mechanism of hot cracking. Figure 3.33a shows a solidification crack formed during LBW of two 5xxx and 6xxx Al alloys in a lap configuration. The crack is initiated in the right side where is close to the free edge. A FE thermal-mechanical analysis of the process helps to interpret the hot crack formation in this region [113]. The temperature distribution obtained by thermal analysis is used to determine where and when the susceptible zone to hot cracking (the mushy zone) forms. Matching these data with the stresses developed at the same time and location gives useful results on the crack formation. The mushy zone that forms in the temperature range of 590–650 °C is highlighted in Figure 3.33b along with the deformation map. The stress distribution in direction of *x*′ is calculated and shown in Figure 3.33c along with the boundary of the mushy zone marked by red dotted line. This direction is perpendicular to the direction of dendrites growth, and hence, the stress in this direction is of importance for hot cracking. A high stress in this direction in the presence of the mushy zone causes hot crack initiation, as marked by the white dotted circle in Figure 3.33c, which is in good agreement with the experimental observation in Figure 3.33a. The applicable result of this simulation is that the increasing the distance of the weld zone to the free edge results in a lower stress and therefore a lower hot crack susceptibility [113].

In butt joining, the propagation of the hot crack is in the middle of the weld where an equiaxed microstructure exists due to high welding speed, while the initiation of hot crack is in the border of solid–liquid (Figure 3.34a) [114]. The simulation results help to understand the mechanism of this trend of hot crack formation. In the middle of the weld, the transverse strain is much higher than the longitudinal strain, which causes the crack to propagate preferentially in the longitudinal direction, despite the presence of equiaxed grains (Figure 3.34b). The longitudinal strain in the BTR zone in the fusion line is very high (Figure 3.34c), which, along with

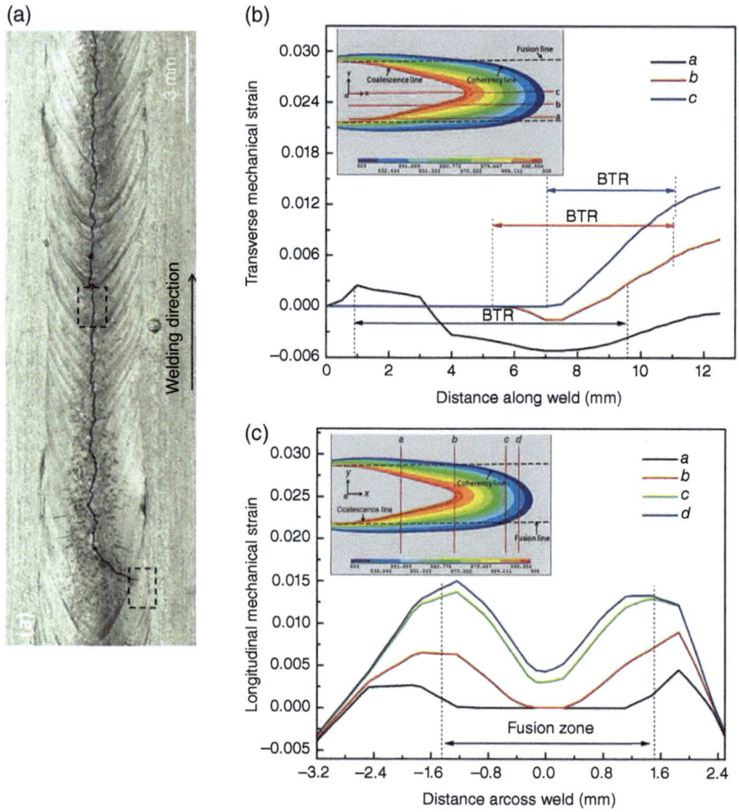

Figure 3.34 (a) Initiation and propagation of hot cracking in laser beam welding of AA6013. (b) Transverse and (c) longitudinal strain along the lines shown, obtained by simulation. Source: Ref. [114]/Sage Publications.

the columnar grains in this region (with a low permeability of liquid to backfill the crack), promotes the initiation of the crack along the columnar grains [114].

The results of thermal modeling can be fed into other simulation methods such as phase field to analyze the evolution of grain structure and solute segregation in the FZ during the simulation. The outputs from the thermal simulation such as the temperature gradient (G) and the solidification interface velocity (R) yield the temperature field at the solid–liquid interface, which are then used in the phase field model to calculate the phases ($\phi(r, t)$) as well as the solute concentration ($c(r, t)$) at any time (t) and location (r) [115]. The results of phase field simulation for laser welding of two 6061/5182 alloys in two different lap configurations are shown in Figure 3.35a–d, showing the phase distribution and the solute concentration, respectively. These results help more accurately to predict the occurrence of SC by knowing that the longer inter-granular channels (observed in Figure 3.35a) hinder the back diffusion to heal the cracks. Furthermore, severe solute segregation in this state (shown in Figure 3.35c) is observed, which aggregates the SC.

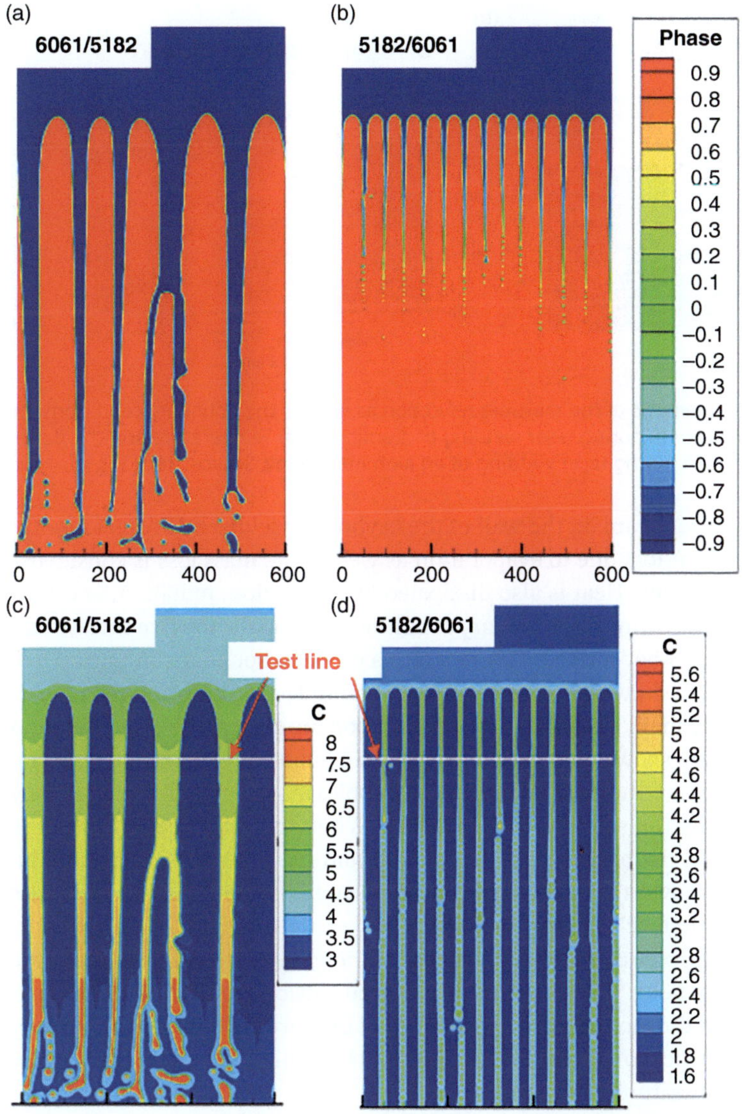

Figure 3.35 (a and b) Phase distribution maps (liquid and solid) at liquid–solid interface during laser welding of 6061/5182 alloys in two different lap configurations (6061 on top and 5182 on top). (c and d) Solute concentration maps [115]. 6061/5182 weld is prone to solidification cracking due to longer inter-granular channels and more severe solute segregation.

The mechanism of pore formation during an unstable keyhole can be understood by simulation. Laser welding is a multi-physical phenomenon, and some simplifications are needed to model it, such as the neglect of the shielding gas as well as the usage of a simple heat source [37]. The forces acting in the keyhole result from the recoil pressure, the buoyancy force, and the surface tension. The multiple

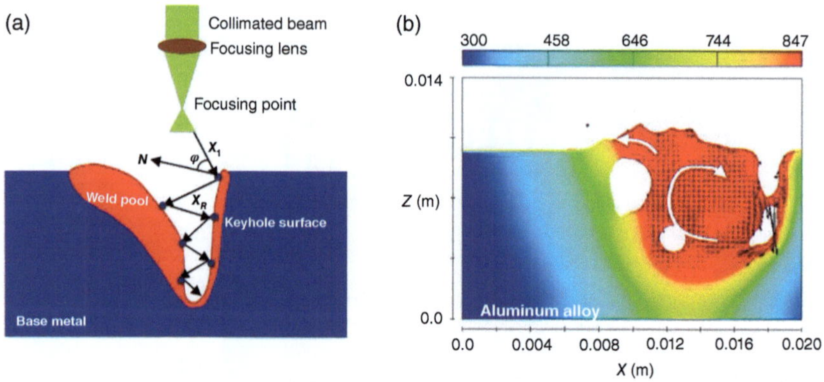

Figure 3.36 (a) Schematic of the multiple reflection as well as the boundary conditions for simulation of the keyhole during laser welding of aluminum. (b) Temperature and the fluid flow field in a keyhole during laser welding of an aluminum alloy. Source: Ref. [37].

reflection of the laser beam on the wall of the keyhole is yielded by Fresnel absorption model. The heat loss due to evaporation as well as the mass loss is considered by the proper equations. Heat is also dissipated by convection, radiation, and heat transfer. There are two kinds of boundary conditions: one on the top free surface and the other one in the other surfaces. There is also a pressure boundary on the top free surface. Figure 3.36a shows the schematic of the multiple reflection and the boundary conditions. The main outcomes of the simulation are temperature and the fluid flow field, which are shown in Figure 3.36b. When a bubble is formed due to unstable keyhole, the gas is entrapped within it. The bubbles move upward as long as they are not captured by the solidification interface. When the solidification interface reaches the bubble, the bubble stops flowing up and the pore forms. The simulation results have shown that the Al solidification rate is high, which makes it less likely for a bubble to reach the surface [37].

The main driving force of material flow is the Marangoni effect, which is due to the surface tension gradient. The source of this surface tension gradient is the temperature gradient and the temperature dependency of the surface tension. The velocity fields in the molten pool of Al obtained by simulation are highest around the keyhole due to a high-temperature gradient [116].

The simulation of the keyhole also helps to realize the mechanism of the spatter formation. The distribution of the vertical momentum and temperature at the top of the keyhole obtained by simulation is shown in Figure 3.37a. Their value is seen to be the highest at the top surface. The surface tension of the melt around the keyhole is the lowest, as the dependency of surface tension to temperature in Al melt is inverse. The force balance around the keyhole is schematically shown in Figure 3.37b. The surface tension (Υ) and gravity (G) forces act in suppression of the spatter while the vapor shear stress (τ_w) and recoil pressure (P_r) enhance it [117].

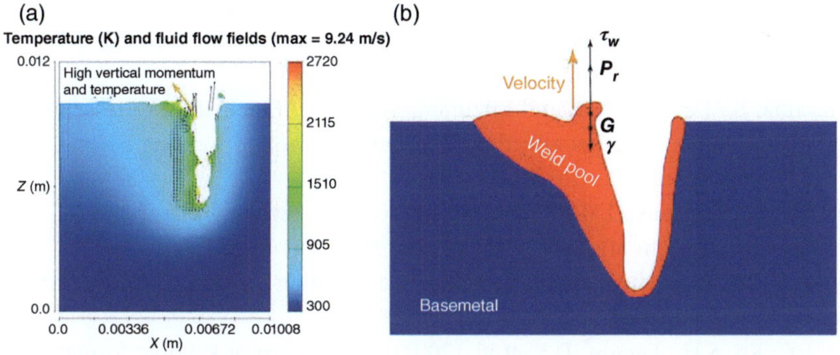

Figure 3.37 (a) Momentum and temperature distribution around the keyhole. (b) Schematic of the force balance around the keyhole in vertical direction. Source: Ref. [117].

References

1 Mirakhorli, F., Nadeau, F., and Guillemette, G. (2018). Single pass laser cold-wire welding of thick section AA6061-T6 aluminum alloy. *Journal of Laser Applications* 30 (3): 032421.

2 Feng, J., Zhang, P., Yan, H. et al. (2023). Application of laser welding in electric vehicle battery manufacturing: a review. *Coatings* 13 (8): 1313.

3 Möller, M. and Vogt, S. (2022). Novel multi focus optic approach for gas-tight welding of aluminum alloys in e-mobility. *Procedia CIRP* 111: 774–777.

4 Xin, X., Xinyong, W., and Juan, L. (2021). Hot-cracking susceptibility and shear fracture behavior of dissimilar Ti_6Al_4V/AA6060 alloys in pulsed Nd : YAG laser welding. *Chinese Journal of Aeronautics* 34 (4): 375–386.

5 Vaidya, W., Horstmann, M., Ventzke, V. et al. (2010). Improving interfacial properties of a laser beam welded dissimilar joint of aluminium AA6056 and titanium Ti6Al4V for aeronautical applications. *Journal of Materials Science* 45: 6242–6254.

6 Enz, J. (2012). Laserstrahlschweissen von hochfesten Aluminium-Lithium-Legierungen. Diploma thesis. Technical University of Hamburg-Harburg.

7 Mahrle, A. and Beyer, E. (2007). Control of the energy deposition during laser beam welding by oscillation techniques. *Proceedings of the Fourth International WLT-Conference on Lasers in Manufacturing*, Munich, Germany (June 2007).

8 Xu, F., Chen, L., He, E.G. et al. (2019). Laser welding 6A02 aluminum alloy with filler wire under high welding speed. *IOP Conference Series: Materials Science and Engineering, Volume 504, the 2nd International Workshop on Materials Science and Mechanical Engineering (IWMSME2018)*, Qingdao, China (26–28 October 2018). IOP Publishing.

9 Cho, W.-I., Schultz, V., and Woizeschke, P. (2018). Numerical study of the effect of the oscillation frequency in buttonhole welding. *Journal of Materials Processing Technology* 261: 202–212.

10 Katayama, S., Kawahito, Y., and Mizutani, M. (2010). Elucidation of laser welding phenomena and factors affecting weld penetration and welding defects. *Physics Procedia* 5: 9–17.

11 Bunaziv, I., Akselsen, O.M., Ren, X. et al. (2021). A review on laser-assisted joining of aluminium alloys to other metals. *Metals* 11 (11): 1680.

12 Volpp, J. (2017). Keyhole stability during laser welding – part II: process pores and spatters. *Production Engineering* 11: 9–18.

13 Lee, J.Y., Ko, S.H., Farson, D.F. et al. (2002). Mechanism of keyhole formation and stability in stationary laser welding. *Journal of Physics D: Applied Physics* 35 (13): 1570.

14 Miyagi, M., Wang, H., Yoshida, R. et al. (2018). Effect of alloy element on weld pool dynamics in laser welding of aluminum alloys. *Scientific Reports* 8 (1): 12944.

15 Haglund, P., Eriksson, I., Powell, J. et al. (2013). Surface tension stabilized laser welding (donut laser welding) – a new laser welding technique. *Journal of Laser Applications* 25 (3): 031501.

16 Schultz, V. and Woizeschke, P. (2018). High seam surface quality in keyhole laser welding: buttonhole welding. *Journal of Manufacturing and Materials Processing* 2 (4): 78.

17 Punzel, E., Hugger, F., Dinkelbach, T. et al. (2020). Influence of power distribution on weld seam quality and geometry in laser beam welding of aluminum alloys. *Procedia CIRP* 94: 601–604.

18 Suttmann, O., Moalem, A., Kling, R. et al. (2010). Drilling, cutting, welding, marking and microforming. In: *Laser Precision Microfabrication* (ed. K. Sugioka, M. Meunier, and A. Piqué), 311–335. Springer.

19 Leong, K., Geyer, H.K., Sabo, K.R. et al. (1997). Threshold laser beam irradiances for melting and welding. *Journal of Laser Applications* 9 (5): 227–231.

20 Tan, C., Xu, B., Liu, F. et al. (2022). Effect of ultrasonic vibration on porosity suppression and columnar-to-equiaxed transition in laser-MIG hybrid welding of aluminum alloy. *The International Journal of Advanced Manufacturing Technology* 122 (5–6): 2463–2474.

21 Von Witzendorff, P., Kaierle, S., Suttmann, O. et al. (2015). In situ observation of solidification conditions in pulsed laser welding of AL6082 aluminum alloys to evaluate their impact on hot cracking susceptibility. *Metallurgical and Materials Transactions A* 46: 1678–1688.

22 Huntington, C. and Eagar, T. (1983). Laser welding of aluminum and aluminum alloys. *Welding Research Supplement* 62 (4): 105.

23 Miah, M.J., Kalosha, V., Rosales, R. et al. (2016). Novel types of photonic band crystal high power and high brightness semiconductor lasers. *Frontiers of Optoelectronics* 9: 225–237.

24 Kashaev, N., Ventzke, V., and Çam, G. (2018). Prospects of laser beam welding and friction stir welding processes for aluminum airframe structural applications. *Journal of Manufacturing Processes* 36: 571–600.

25 Timmermann, A., Meinschien, J., Bruns, P. et al. (2008). Next generation high-brightness diode lasers offer new industrial applications. *Proceedings of SPIE 6876, in High-Power Diode Laser Technology and Applications VI*, San Jose, CA (25 February 2008). SPIE.

26 Ming, G., Xiaoyan, Z., and Qianwu, H. (2007). Effects of gas shielding parameters on weld penetration of CO_2 laser-TIG hybrid welding. *Journal of Materials Processing Technology* 184 (1–3, 183): 177.

27 Xiao, R. and Zhang, X. (2014). Problems and issues in laser beam welding of aluminum–lithium alloys. *Journal of Manufacturing Processes* 16 (2): 166–175.

28 Gao, M., Chen, C., Hu, M. et al. (2015). Characteristics of plasma plume in fiber laser welding of aluminum alloy. *Applied Surface Science* 326: 181–186.

29 Wei, P. and Chao, T. (2014). Prediction of pore size in high power density beam welding. *International Journal of Heat and Mass Transfer* 79: 223–232.

30 Kawahito, Y., Mizutani, M., and Katayama, S. (2008). Optical interaction between laser beam and induced plume in the ultra-high power density fiber laser welding of stainless steel. *Transactions of Joining and Welding Research Institute* 37 (2): 19–25.

31 Enz, J., Riekehr, S., Ventzke, V. et al. (2016). Fibre laser welding of high-alloyed Al–Zn–Mg–Cu alloys. *Journal of Materials Processing Technology* 237: 155–162.

32 Miyagi, M., Kawahito, Y., Kawakami, H. et al. (2017). Dynamics of solid-liquid interface and porosity formation determined through x-ray phase-contrast in laser welding of pure Al. *Journal of Materials Processing Technology* 250: 9–15.

33 Yan, S., Zhu, Z., Ma, C. et al. (2019). Porosity formation and its effect on the properties of hybrid laser welded Al alloy joints. *The International Journal of Advanced Manufacturing Technology* 104: 2645–2656.

34 Huang, L., Hua, X., Wu, D. et al. (2018). Effect of magnesium content on keyhole-induced porosity formation and distribution in aluminum alloys laser welding. *Journal of Manufacturing Processes* 33: 43–53.

35 Pastor, M., Zhao, H., and Debroy, T. (2001). Pore formation during continuous wave Nd: YAG laser welding of aluminium for automotive applications. *Welding International* 15 (4): 275–281.

36 Deng, A., Chen, H., Zhang, Y. et al. (2023). Effect of filler materials on the porosity formation of aluminum alloy by laser welding with filler wire. *Optics & Laser Technology* 159: 109000.

37 Huang, L., Hua, X., Wu, D. et al. (2018). Numerical study of keyhole instability and porosity formation mechanism in laser welding of aluminum alloy and steel. *Journal of Materials Processing Technology* 252: 421–431.

38 Geng, S., Yang, W., Jiang, P. et al. (2022). Numerical study of keyhole dynamics and porosity formation during high-power oscillating laser welding of medium-thick aluminum alloy plates. *International Journal of Heat and Mass Transfer* 194: 123084.

39 Huang, L., Hua, X., Wu, D. et al. (2019). Experimental investigation and numerical study on the elimination of porosity in aluminum alloy laser welding and laser–GMA welding. *Journal of Materials Engineering and Performance* 28: 1618–1627.

40 Volpp, J., Srowig, J., and Vollertsen, F. (2016). Spatters during laser deep penetration welding with a bifocal optic. *Advanced Materials Research* 1140: 123–129.

41 Bielenin, M. (2021). *Prozessstrategien zur Vermeidung von Heißrissen beim Schweißen von Aluminium mit pulsmodulierbaren Laserstrahlquellen.* BoD–Books on Demand.

42 Hagenlocher, C., Weller, D., Weber, R. et al. (2019). Analytical description of the influence of the welding parameters on the hot cracking susceptibility of laser beam welds in aluminum alloys. *Metallurgical and Materials Transactions A* 50: 5174–5180.

43 Miyagi, M., Kawahito, Y., Wang, H. et al. (2018). X-ray phase contrast observation of solidification and hot crack propagation in laser spot welding of aluminum alloy. *Optics Express* 26 (18): 22626–22636.

44 Wang, X., Wang, H.-P., Lu, F. et al. (2014). Analysis of solidification cracking susceptibility in side-by-side dual-beam laser welding of aluminum alloys. *The International Journal of Advanced Manufacturing Technology* 73: 73–85.

45 Rakhi, K., Kang, S., and Shin, J. (2023). Hot-cracking mechanism of laser welding of aluminum alloy 6061 in lap joint configuration. *Materials (Basel)* 16 (19): 6426.

46 Hekmatjou, H. and Naffakh-Moosavy, H. (2018). Hot cracking in pulsed Nd: YAG laser welding of AA5456. *Optics & Laser Technology* 103: 22–32.

47 Tang, Z. and Vollertsen, F. (2014). Influence of grain refinement on hot cracking in laser welding of aluminum. *Welding in the World* 58: 355–366.

48 Wang, J., Wang, H.-P., Wang, X. et al. (2015). Statistical analysis of process parameters to eliminate hot cracking of fiber laser welded aluminum alloy. *Optics & Laser Technology* 66: 15–21.

49 Zhang, Y., Lu, F., Cui, H. et al. (2016). Investigation on the effects of parameters on hot cracking and tensile shear strength of overlap joint in laser welding dissimilar Al alloys. *The International Journal of Advanced Manufacturing Technology* 86: 2895–2904.

50 Zhang, Y., Lu, F., Wang, H.-P. et al. (2015). Reduced hot cracking susceptibility by controlling the fusion ratio in laser welding of dissimilar Al alloys joints. *Journal of Materials Research* 30 (7): 993–1001.

51 Enz, J., Carrarin, C., Riekehr, S. et al. (2018). Hot cracking behaviour of an autogenously laser welded Al--Cu–Li alloy. *The International Journal of Advanced Manufacturing Technology* 95: 299–310.

52 Norouzian, M., Elahi, M.A., and Plapper, P. (2023). A review: suppression of the solidification cracks in the laser welding process by controlling the grain structure and chemical compositions. *Journal of Advanced Joining Processes* 7 (9): 100139.

53 Tang, Z., Seefeld, T., and Vollertsen, F. (2011). Grain refinement by laser welding of AA 5083 with addition of Ti/B. *Physics Procedia* 12: 123–133.

54 Holzer, M., Hofmann, K., Mann, V. et al. (2016). Change of hot cracking susceptibility in welding of high strength aluminum alloy AA 7075. *Physics Procedia* 83: 463–471.

55 Vyskoč, M., Sahul, M., and Sahul, M. (2018). Effect of shielding gas on the properties of AW 5083 aluminum alloy laser weld joints. *Journal of Materials Engineering and Performance* 27: 2993–3006.

56 Vyskoč, M., Dománková, M., Sahul, M. et al. (2020). The effect of the shielding gas flow rate on the geometry, porosity, microstructure and mechanical properties of laser weld joints. *Research Papers Faculty of Materials Science and Technology Slovak University of Technology* 28 (47): 72–85.

57 Bunaziv, I., Akselsen, O.M., Ren, X. et al. (2021). Laser beam and laser-arc hybrid welding of aluminium alloys. *Metals* 11 (8): 1150.

58 Li, S., Zhang, L.-J., Ning, J. et al. (2020). Comparative study on the microstructures and properties of wire+ arc additively manufactured 5356 aluminium alloy with argon and nitrogen as the shielding gas. *Additive Manufacturing* 34: 101206.

59 Cai, C., He, S., Chen, H. et al. (2019). The influences of Ar–He shielding gas mixture on welding characteristics of fiber laser-MIG hybrid welding of aluminum alloy. *Optics & Laser Technology* 113: 37–45.

60 Ahn, J., He, E., Chen, L. et al. (2017). The effect of Ar and He shielding gas on fibre laser weld shape and microstructure in AA 2024-T3. *Journal of Manufacturing Processes* 29: 62–73.

61 Xia, H., Tan, C., Ruyu, T. et al. (2020). Influence of shielding gas on microstructure and mechanical properties of laser welded–brazed Al/steel lapped joint. *Journal of Manufacturing Processes* 54: 347–358.

62 Atabaki, M.M., Yazdian, N., and Kovacevic, R. (2016). Partial penetration laser-based welding of aluminum alloy (AA 5083-H32). *Optik* 127 (16): 6782–6804.

63 Wang, J., Genchen, P., Li, L.Q. et al. (2019). 30 kW-level laser welding characteristics of 5A06 aluminum alloy thick plate under subatmospheric pressure. *Optics & Laser Technology* 119: 105668.

64 Jiang, M., Chen, X., Chen, Y. et al. (2020). Mitigation of porosity defects in fiber laser welding under low vacuum. *Journal of Materials Processing Technology* 276: 116385.

65 Lee, Y., Cheon, J., and Kang, M. (2021). Vacuum laser beam welding characteristics for aluminum alloys and its effects on plasma plume generation and laser weldability. *The International Journal of Advanced Manufacturing Technology* 115 (1): 531–539.

66 Bergmann, J.P., Bielenin, M., Stambke, M. et al. (2013). Effects of diode laser superposition on pulsed laser welding of aluminum. *Physics Procedia* 41: 180–189.

67 Volpp, J. and Vollertsen, F. (2019). Impact of multi-focus beam shaping on the process stability. *Optics & Laser Technology* 112: 278–283.

68 Tzeng, Y.-F. (2000). Process characterisation of pulsed Nd: YAG laser seam welding. *The International Journal of Advanced Manufacturing Technology* 16 (1): 10–18.

69 Haddad, E., Chung, W.S., Katz, O. et al. (2022). Laser micro welding with fiber lasers for battery and fuel cell based electromobility. *Journal of Advanced Joining Processes* 5: 100085.

70 Sanati, S., Nabavi, S.F., Esmaili, R. et al. (2024). A comprehensive review of laser wobble welding processes in metal materials: processing parameters and practical applications. *Lasers in Manufacturing and Materials Processing* 1–37.

71 Rohde, M., Markert, C., and Pfleging, W. (2010). Laser micro-welding of aluminum alloys: experimental studies and numerical modeling. *The International Journal of Advanced Manufacturing Technology* 50: 207–215.

72 Von Witzendorff, P., Hermsdorf, J., Kaierle, S. et al. (2015). Double pulse laser welding of 6082 aluminium alloys. *Science and Technology of Welding and Joining* 20 (1): 42–47.

73 Kotalik, P. and Boeck, T. (2008). Modelling of heat transfer and fluid flow in laser welding. *PAMM: Proceedings in Applied Mathematics and Mechanics* 8 (1): 10623–10624. Wiley Online Library.

74 Yan, S., Nie, Y., Zhu, Z. et al. (2014). Characteristics of microstructure and fatigue resistance of hybrid fiber laser-MIG welded Al–Mg alloy joints. *Applied Surface Science* 298: 12–18.

75 Zhang, C., Li, G., Gao, M. et al. (2013). Microstructure and process characterization of laser-cold metal transfer hybrid welding of AA6061 aluminum alloy. *The International Journal of Advanced Manufacturing Technology* 68: 1253–1260.

76 Peleshenko, S., Viktor, K., Khaskin, V. et al. (2022). Features of physical and metallurgical processes during welding of thin-walled aluminum alloy structures using laser radiation. *Danish Scientific Journal* 65: 50–59.

77 Wang, L., Gao, M., Zhang, C. et al. (2016). Effect of beam oscillating pattern on weld characterization of laser welding of AA6061-T6 aluminum alloy. *Materials & Design* 108: 707–717.

78 Wang, Z., Oliveira, J.P., Zeng, Z. et al. (2019). Laser beam oscillating welding of 5A06 aluminum alloys: microstructure, porosity and mechanical properties. *Optics & Laser Technology* 111: 58–65.

79 Wu, Q., Xiao, R.S., Zou, J.L. et al. (2018). Weld formation mechanism during fiber laser welding of aluminum alloys with focus rotation and vertical oscillation. *Journal of Manufacturing Processes* 36: 149–154.

80 Nothdurft, S., Ohrdes, H., Twiefel, J. et al. (2019). Influence of ultrasonic amplitude and position in the vibration distribution on the microstructure of a laser beam welded aluminum alloy. *Journal of Laser Applications* 31 (2): 022402.

81 Ohrdes, H., Seffer, S., Nowroth, C. et al. (2021). Influence of the ultrasonic vibration amplitude on the melt pool dynamics and the weld shape of laser beam welded EN AW-6082 utilizing a new excitation system for laser beam welding. *Production Engineering* 15: 151–160.

82 Völkers, S., Scharifi, E., Sajjadifar, S.V. et al. (2020). On the influence of in situ sound wave superposition on the microstructure of laser welded 7000 aluminum alloys. *Journal of Advanced Joining Processes* 1: 100013.

83 Chen, Q., Ge, H., Yang, C. et al. (2017). Study on pores in ultrasonic-assisted TIG weld of aluminum alloy. *Metals* 7 (2): 53.

84 Liu, J., Zhu, H., Li, Z. et al. (2019). Effect of ultrasonic power on porosity, microstructure, mechanical properties of the aluminum alloy joint by ultrasonic assisted laser-MIG hybrid welding. *Optics & Laser Technology* 119: 105619.

85 Liverani, E., Ascari, A., and Fortunato, A. (2023). The role of filler wire and scanning strategy in laser welding of difficult-to-weld aluminum alloys. *The International Journal of Advanced Manufacturing Technology* 128 (1–2): 763–777.

86 Park, Y., Yu, J., and Rhee, S. (2010). A study on the weld characteristics of 5182 aluminum alloy by Nd: YAG laser welding with filler wire for car bodies. *International Journal of Automotive Technology* 11: 729–736.

87 Huang, W., Xiao, J., Chen, S.S. et al. (2022). Control of wire melting behavior during laser hot wire deposition of aluminum alloy. *Optics & Laser Technology* 150: 107978.

88 Knapp, G.L., Gussev, M., Shyam, A. et al. (2022). Microstructure, deformation and fracture mechanisms in Al-4043 alloy produced by laser hot-wire additive manufacturing. *Additive Manufacturing* 59: 103150.

89 Schwarz, A., Schleser, M., Gerhards, B. et al. (2021). Welding of additive manufactured Alsi10mg: using laser welding in a vacuum for high quality weld seams – a new approach to welding LPBF manufactured $AlSi_{10}Mg$. *South African Journal of Industrial Engineering* 32 (4): 99–112.

90 Fetzer, F., Sommer, M., Weber, R. et al. (2018). Reduction of pores by means of laser beam oscillation during remote welding of AlMgSi. *Optics and Lasers in Engineering* 108: 68–77.

91 Dittrich, D., Jahn, A., Standfuss, J. et al. (2017). Laser beam welding of atmosphere aluminum die cast material using high frequency beam oscillation and brilliant beam sources. *Journal of Laser Applications* 29 (2).

92 Epperlein, M., Schiebahn, A., Reisgen, U. et al. (2024). Joining aluminum die castings and wrought aluminum by resistance spot welding. *Journal of Materials: Design and Applications* 238 (12): 14644207241236388.

93 Fritzsche, A., Hilgenberg, H., Teichmann, F. et al. (2018). Improved degassing in laser beam welding of aluminum die casting by an electromagnetic field. *Journal of Materials Processing Technology* 253: 51–56.

94 Yu, X., Fan, D., Huang, J. et al. (2019). Arc-assisted laser welding brazing of aluminum to steel. *Metals* 9 (4): 397.

95 Wang, J., Fu, X., Zhang, L. et al. (2021). A short review on laser welding/brazing of aluminum alloy to steel. *International Journal of Advanced Manufacturing Technology* 112: 2399–2411.

96 Guan, Q., Jiangqi, L., Ping, Y. et al. (2019). Effect of steel to aluminum laser welding parameters on mechanical properties of weld beads. *Optics & Laser Technology* 111: 387–394.

97 Hollatz, S., Heinen, P., Limpert, E. et al. (2020). Overlap joining of aluminium and copper using laser micro welding with spatial power modulation. *Welding in the World* 64: 513–522.

98 Solchenbach, T., Plapper, P., and Cai, W. (2014). Electrical performance of laser braze-welded aluminum–copper interconnects. *Journal of Manufacturing Processes* 16 (2): 183–189.

99 Alikhani, A., Beygi, R., Mehrizi, M.Z. et al. (2021). Effect of Mg and Si on intermetallic formation and fracture behavior of pure aluminum-galvanized carbon-steel joints made by weld-brazing. *Journal of Central South Univedrsity* 28 (11): 3626–3638.

100 Dharmendra, C., Rao, K.P., Wilden, J. et al. (2011). Study on laser welding–brazing of zinc coated steel to aluminum alloy with a zinc based filler. *Materials Science and Engineering: A* 528 (3): 1497–1503.

101 Khayoon, M.A., Hubeatir, K.A., and Al-Khafaji, M.M. (2021). Laser transmission welding is a promising joining technology technique – a recent review. *Journal of Physics: Conference Series* 1973: 012023. IOP Publishing.

102 Goyal, D.K. and Kant, R. (2024). Mechanical and morphological analysis of laser transmission welded dissimilar plastics using metal absorber. *Optics & Laser Technology* 170: 110238.

103 Min, Z., Yufei, C., Changjun, C. et al. (2021). A new sealing technology for ultra-thin glass to aluminum alloy by laser transmission welding method. *International Journal of Advanced Manufacturing Technology* 115 (7–8): 2017–2035.

104 Bideskan, A.S., Ebrahimzadeh, P., and Teimouri, R. (2020). Fabrication of bi-layer PMMA and aluminum 6061-T6 laminates by laser transmission welding: performance prediction and optimization. *International Journal of Lightweight Materials and Manufacture* 3 (2): 150–159.

105 Devrient, M., Frick, T., and Schmidt, M. (2011). Laser transmission welding of optical transparent thermoplastics. *Physics Procedia* 12: 157–165.

106 Gao, S., Geng, S., Jiang, P. et al. (2022). Numerical study on the effect of residual stress on mechanical properties of laser welds of aluminum alloy 2024. *Optics & Laser Technology* 146: 107580.

107 Evdokimov, A., Doynov, N., Ossenbrink, R. et al. (2021). Thermomechanical laser welding simulation of dissimilar steel-aluminum overlap joints. *International Journal of Mechanical Sciences* 190: 106019.

108 Huang, Z., Luo, Z., Ao, S. et al. (2018). Effect of laser welding parameters on weld bowing distortion of thin plates. *High Temperature Materials and Processes* 37 (4): 299–311.

109 Chuang, T.-C., Lo, Y.-L., Tran, H.-C. et al. (2023). Optimization of butt-joint laser welding parameters for elimination of angular distortion using high-fidelity simulations and machine learning. *Optics & Laser Technology* 167: 109566.

110 Wei, H., Chen, J.S., Wang, H.-P. et al. (2016). Thermomechanical numerical analysis of hot cracking during laser welding of 6XXX aluminum alloys. *Journal of Laser Applications* 28 (2): 022405.

111 Beygi, R., Marques, E., and da Silva, L.F. (2022). *Computational Concepts in Simulation of Welding Processes*. Springer.

112 Lei, Z., Zou, J., Wang, D. et al. (2020). Finite-element inverse analysis of residual stress for laser welding based on a contour method. *Optics & Laser Technology* 129: 106289.

113 Wei, H., He, Q., Chen, J.-S. et al. (2017). Coupled thermal-mechanical-contact analysis of hot cracking in laser welded lap joints. *Journal of Laser Applications* 29 (2): 022412.

114 Wang, X., Lu, F.G., Wang, H.P. et al. (2015). Experimental and numerical analysis of solidification cracking behaviour in fibre laser welding of 6013 aluminium alloy. *Science and Technology of Welding and Joining* 20 (1): 58–67.

115 Chen, L., Wang, C., Xiong, L. et al. (2020). Microstructural, porosity and mechanical properties of lap joint laser welding for 5182 and 6061 dissimilar aluminum alloys under different place configurations. *Materials & Design* 191: 108625.

116 Geng, S., Jiang, P., Shao, X. et al. (2020). Heat transfer and fluid flow and their effects on the solidification microstructure in full-penetration laser welding of aluminum sheet. *Journal of Materials Science & Technology* 46: 50–63.

117 Wu, D., Ye, Y., Huang, L. et al. (2018). Experimental and numerical study of spatter formation and composition change in fiber laser welding of aluminum alloy. *Journal of Physics D: Applied Physics* 51 (18): 185604.

4

Resistance Welding Process

4.1 Introduction

RSW is based on heat generation from the electrical resistivity of the contact surfaces of two adjoints. It does not need flux, filler material, and protective gases and is well suited for mass production. This process uses two electrodes pressing the outer surfaces of two materials, through which electricity passes and produces heat at the interface. The cycle of pressure and heat and their magnitudes are the general parameters of this process. The schematic of this process and the cycles of pressure and current are shown Figure 4.1a and b, respectively. This process is a fusion-based process, though the fusion region is invisible. The extent of the nugget formed depends on the electrode size and the heat generated (H) which is obtained from [1]:

$$H = RI^2t \tag{4.1}$$

where R is the electrical resistance, I is current, and t is time.

DC, medium-frequency direct current (MFDC), and alternative current (AC) are used to join Al alloys by RSW. The MFDC is more energy-efficient to produce the same nugget size than AC [2]. When using DC current, the Peltier effect causes the nugget formation to be different at the top and bottom. When a current goes through the joint of two materials with different conductivities, the heat is generated or absorbed depending on the current direction. The electrode made of copper in contact with Al makes a high difference in contact potential at the junction, which causes the Peltier effect to manifest itself more. When the electrode tip is positive, the heat is generated, and when the electrode tip is negative, the heat is absorbed. So, the nugget on the side of the positive electrode is larger [3, 4]. The guidelines for adjustment of the parameters of RSW of Al alloys are mainly provided for wrought alloys, such as DIN EN ISO 18595, DVS 2932, VDA 238-401, and AWS D8.2M:2017 [5]. For the welding lobe determination, a graph is plotted (Figure 4.2a), whose vertical axis is the diameter of the spot and the horizontal one is the main current. Three regions are identified in this graph: region I where $d_{spot} < 4\sqrt{t}$, region II where $4\sqrt{t} < d_{spot} < 5\sqrt{t}$, and region III where $5\sqrt{t} < d_{spot}$ (t is the thickness of the sheet).

Welding Metallurgy of Aluminium Alloys: Design, Processes, and Simulations, First Edition.
Reza Beygi, Eduardo André de Sousa Marques, Ricardo João Camilo Carbas, and Lucas Filipe Martins da Silva.

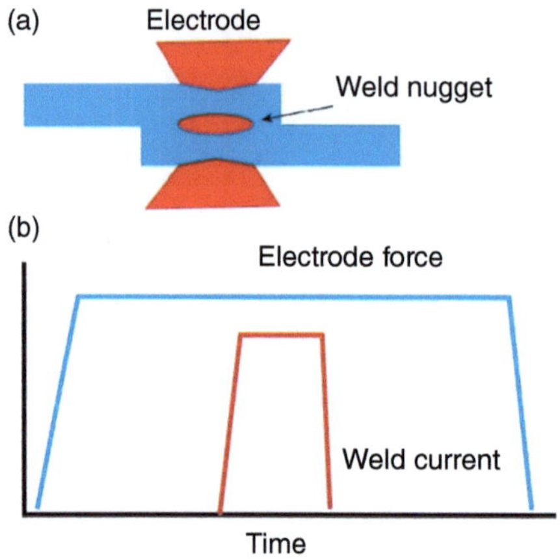

Figure 4.1 (a) Schematic of resistance spot welding. (b) Cycles of load and current. Source: Ref. [1].

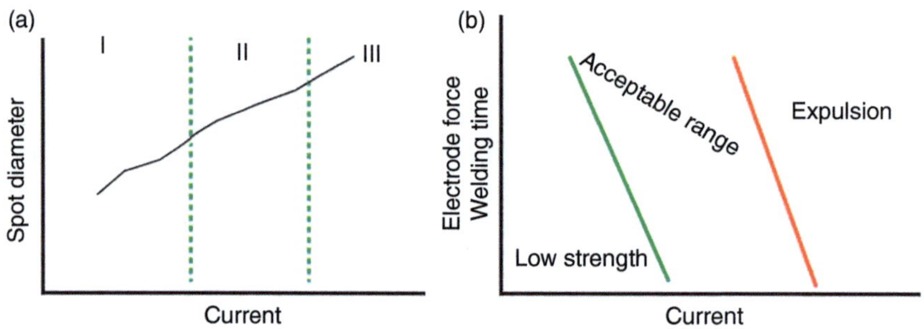

Figure 4.2 (a) Schematic of welding lobe determination for RSW of Al alloy according to VDA 238-401. (b) Process window for obtaining acceptable parameters.

According to VDA 238-401 there is a limit to the maximum current beyond which the expulsion occurs. Taking into account all the main parameters of RSW, a process window can be obtained as depicted in Figure 4.2b.

RSW is mostly used for joining steel and is rarely used for Al. The heat conduction in Al is very high, and its electrical resistance is very low compared to steel. Therefore, a shorter time and a higher current are needed to join them with resistance welding (RW), as the heat is transmitted away rapidly by conduction. The evolution of the dynamic resistance during RSW of Al is different from that of steel. The dynamic resistance of steel starts to decrease and then increase to a peak and eventually decreases to a stable resistance. The rise of the dynamic resistance of steel is due to an increase in the resistance of the bulk caused by a rise in temperature. This increase does not exist during RSW of Al due to its low electrical resistance (Figure 4.3) [6]. When the workpiece is melted, the resistance reaches its minimum. Due to the absence of the peak in the dynamic resistance, the nugget is smaller in Al than in steel.

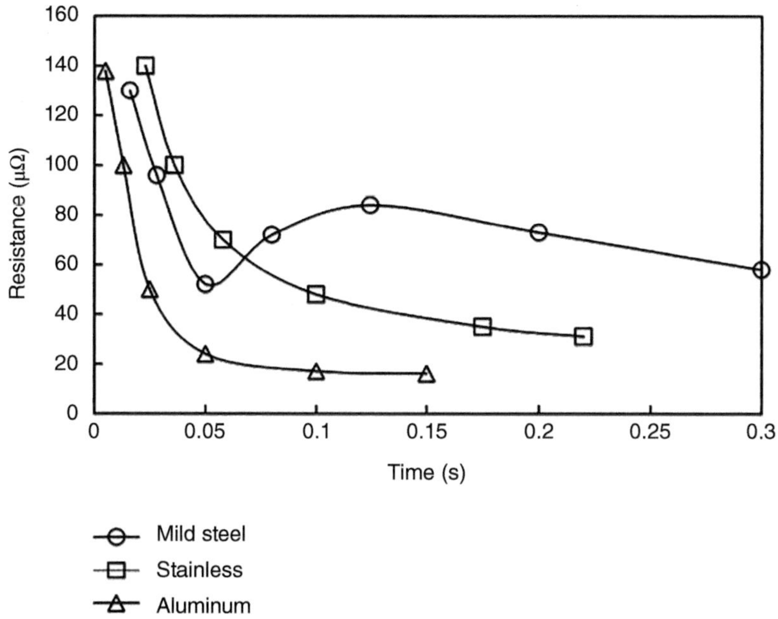

Figure 4.3 Comparison of evolution of dynamic resistance during resistance spot welding of Al and steel. Source: Ref. [7].

Al has an invisible oxide layer with a high melting temperature that adheres to its surface. Even after removal by chemical or mechanical methods, this oxide layer reforms quickly. Issues arising from this oxide layer affect both the electrode/workpiece and workpiece/workpiece interfaces. This oxide layer makes a high electrical resistance at the surface contact of the electrode and Al, causing Al to stick to the electrode during welding. This adds an additional task of removing the stuck Al from the surface of the electrode [8]. A full removing of the oxide layer before the process helps to avoid this problem [9]. The cycle of an electrode life for RSW of Al is shown in Figure 4.4, which correlates the joint strength with the contact area between the electrode and workpiece. In stage I, the contact area increases, leading to an increase in joint strength. In stage II, pitting begins, and in stage III, it expands, reducing the contact area. By stage IV, large cavities form on the electrode, and the joint strength reaches its failure point [10].

The electrodes for Al are usually cambered point surfaces and need to be cleaned due to contaminations. This cleaning is necessary to reduce the resistance of the contact area. The surface of Al also needs cleaning to remove the oxide layer. Shot blasting and pickling in an aqueous solution of 60 g of NaOH per liter water is one technique to remove the oxide layer [12]. During the pickling, the oxide layer or hydroxides dissolve in the solution by stirring. After pickling, the surface is rinsed in the hot water and ethanol. Shot blasting is more effective in reducing the resistance of the contact area, as it produces roughness on the surface of Al, which

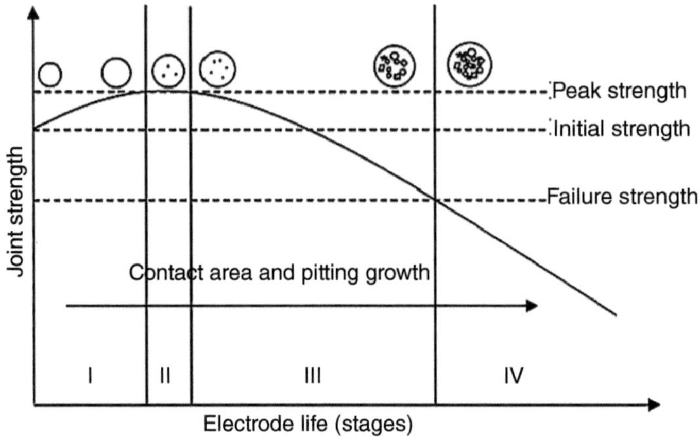

Figure 4.4 Schematic of electrode life used for resistance welding of Al. Source: Ref. [11].

deforms during RSW and breaks the oxide layer. It is suggested to shot-blast only one side of the Al sheets in contact with the electrodes to decrease as much as possible the electrical resistance at the interface of Al/electrode [12]. The combination of aqueous alkali and acid solution also removes the oxide layer effectively. The chemical reactions that lead to the removal of the oxide layers and hydroxide layers are [13]:

$$Al_2O_3 + 2NaOH + 3H_2O \rightarrow 2Na[Al(OH)_4] \quad (4.2)$$

$$Al(OH)_3 + NaOH \rightarrow Na[Al(OH)_4] \quad (4.3)$$

$Al(OH)_3$ is sticky and forms repeatedly, which can be eliminated by acid washing [13]:

$$Al(OH)_3 + 3HNO_3 \rightarrow Al(NO_3)_3 + 3H_2O \quad (4.4)$$

Other methods such as etching with HF solution are also effective in removing the oxide layer [14]. The oxide layer at the interface of two Al workpieces increases the electrical contact resistance when the thickness of the oxide layer is high. In this case, the current is shunted in the weld and a scatter of contact resistance at the interface is observed. The excessive resistance at local points causes an overheating and expulsion. During the expulsion, the molten metal is ejected from the FZ and hence the nugget collapses [15]. One way to overcome this issue is to increase the electrode force to disrupt the oxide layer by local deformation at the interface. But this method is not effective, as it cannot remove the oxide layer completely. A preheating current helps to decrease the contact resistance and make it regular and stable. The current cycle of preheating is shown in Figure 4.5. The mechanism of this improvement is production of local melting in the conducting spots during the application of preheating current, which expands the contact area for the subsequent application of the main current [16]. The schedule of this procedure is shown in Figure 4.5. The surface oxide poses greater challenges during RSW of pure Al than Al alloys, as pure Al is more prone to thick oxide layer formation [12].

Figure 4.5 Schedule of applying preheating (current I_1 at interval time of t_1) before the main welding cycle (current I_2 at time t_3).

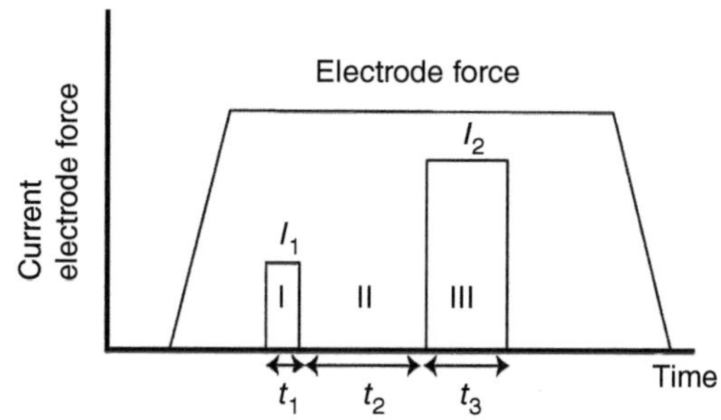

The end of life of an electrode occurs when it produces the nugget sizes below a critical size or when the tensile shear strength of the joint made by it is lowered to a certain percentage of its initial value. The main reason for electrode degradation is alloying of it with Al, which takes place at high temperatures during the RSW process. This alloying is pronounced in some regions more than the others, due to the nonuniformity of the Al oxide layer on the surface of Al. In the regions where the thickness of the oxide layer is lower, higher current passes, which causes pitting to occur on the electrode due to alloying (Figure 4.6) [17]. When the thickness of the oxide layer is low, it fractures more uniformly and provides a large amount of contact spots, avoiding very localized pitting. The AC is advantageous to DC or MFDC, as in the latter cases the positive electrode possesses higher temperature due to Peltier effect, which causes a nonuniform failure of two electrodes. In AC, both electrodes degrade uniformly, resulting in a higher lifespan of the electrodes [18]. Cladding of Al adjoints with another type of Al, which improves the weldability by lowering less expulsion, pores and cracks, can improve the electrode life by extending the welding lobe and lowering the necessary current to make the same nugget size [19]. Using a lubricant between the workpiece and the electrode is another way to extend the electrode life. The lubricant reduces the thickness of the oxide layer, which in turn lowers resistance and heat production, thereby reducing the alloying of the electrode with Al and minimizing pitting [10].

Figure 4.6 Mechanism of electrode pitting during resistance spot welding of Al. (a) Al pick-up. (b) Al–Cu alloying. (c) Electrode pitting. (d) Cavitation.
Source: Ref. [11].

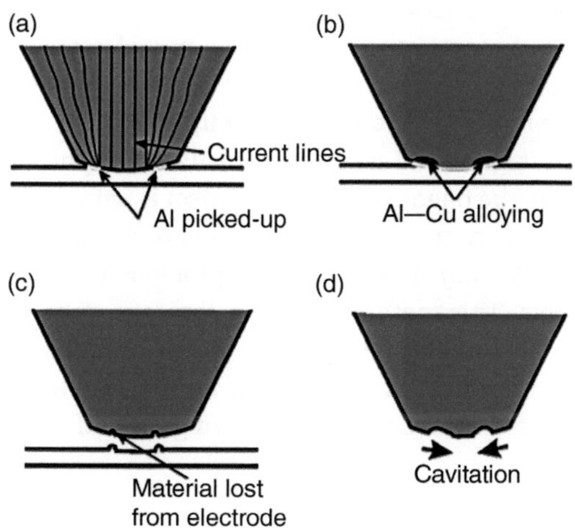

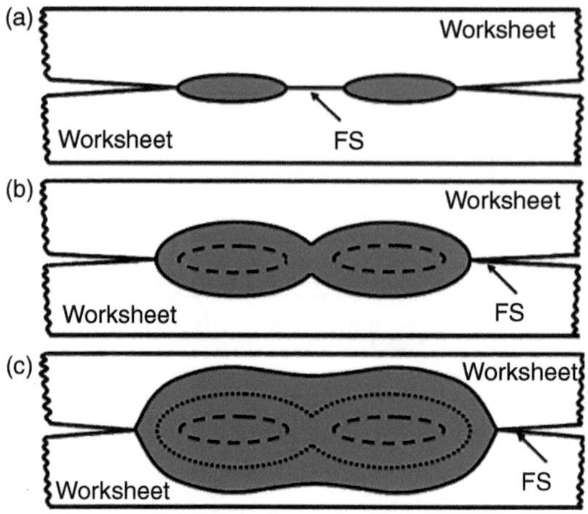

Figure 4.7 Mechanism of nugget formation during resistance spot welding of a 5xxx Al alloy using a flat electrode. Source: Ref. [20].

The shape of the electrode determines the way the melt is produced at the interface of two sheets. A flat surface of the electrode causes a high pressure at the periphery of the electrode, causing the oxide layer to break first around the nugget. This makes the periphery to be the first place to melt. Further application of current and pressure causes this doughnut shape to move toward the center zone, completing the nugget formation (Figure 4.7) [20].

The best shape of electrode used for RSW of Al is a dome shape, whose radius of curvature is optimized to break the oxide layers during the process. Multi-ring dome-shape electrode geometry, developed by General Motors, significantly improved the weldability of Al by RSW [21]. These concentric rings help to remove the oxide layer on the surface of Al by deformation. The textures on the surface of the electrodes also help to break the oxide layer on the surface of Al, which can increase the electrode life. Besides texture and geometry, the coating of the electrodes also affects the life and productivity of the electrodes, but a straightforward result was not observed by this method. This is because the coating, though it strengthens the electrode surface, produces higher heat at the interface [11]. The material of electrode is also decisive in its life. Some electrodes are made of copper reinforced with Al_2O_3 particles to raise the hardness and the softening temperature [22]. Some others are alloyed such as CuCrZr to strengthen it and improve the softening temperature [12].

When several spot welds are made, the preexisting welds divert the electric current from the ongoing weld. This effect is named shunting effect. This causes an insufficient current from the ongoing spot weld, which leads to a small size of the nugget (Figure 4.8). To alleviate this problem, a minimum distance is needed between the spots. This method works in the case of RSW of steel, but in the case of Al, it is more complicated as Al has a high conductivity. A larger distance is needed for Al to solve this issue, but this is not applicable due to structural limitations. The best way to overcome shunting during RSW of Al is to increase the current [6, 23].

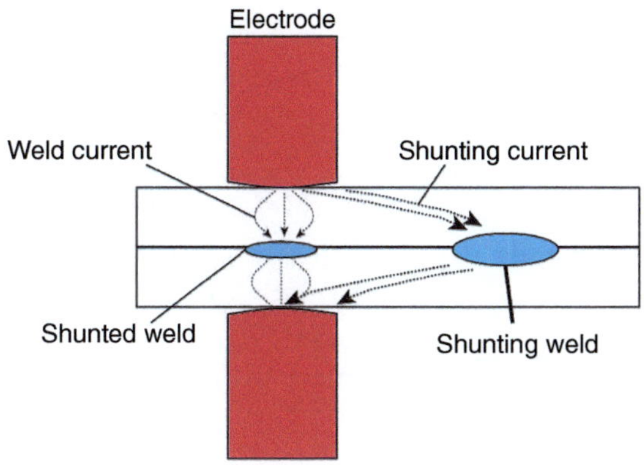

Figure 4.8 Schematic of the shunting effect during RSW.

4.2 Weld Zones

4.2.1 Nugget Zone

The nugget zone has a solidification structure consisting of various microstructures depending on the solidification rate (R) and thermal gradient (G) (Figure 4.9) [24].

The FZ of heat-treatable Al alloys such as 6XXX usually is in solid-solution state, meaning that the precipitates are dissolved. Hence, the lowest hardness for these alloys is observed in the FZ [14]. The solidification redistributes the solutes, which influences the age-hardening behavior. The effective solutes such as Mg and Si in AA6061 segregate at inter-dendritic regions of the FZ, lowering the degree of supersaturation inside the dendrites. Hence, a PWHT only consisting of aging cannot retrieve the hardness in the FZ. To retrieve the hardness in the FZ, a solution treatment before the age-hardening is necessary to homogenize the distribution of solutes. This heat treatment may be performed in the furnace [24].

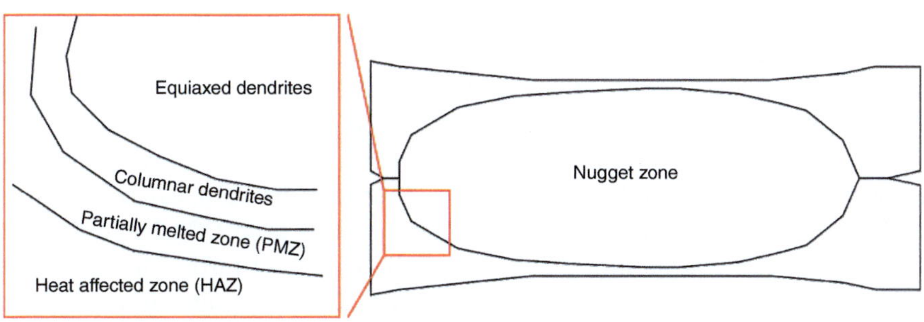

Figure 4.9 Various microstructures observed in the different zones of the weld of Al alloy welded by resistance spot welding.

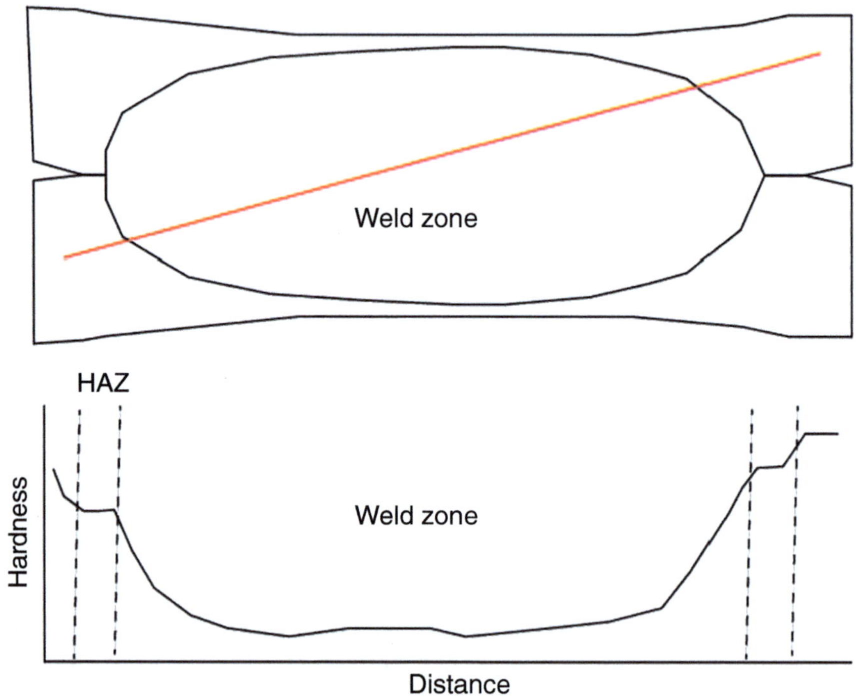

Figure 4.10 Schematic of microhardness profile in a 5xxx Al alloy welded by resistance spot welding.

In the case of non-heat-treatable Al alloys such as 5XXX, a similar trend in the hardness of the nugget zone is observed, as can be seen in Figure 4.10. The decline of hardness in the FZ of these Al alloys is mainly due to the casting microstructure of the FZ. This causes an absence of the hardening mechanism, which exists in the base metal, such as strain hardening [25].

The morphology of the grains in the FZ of Al welded by RSW depends on G and R values, as mentioned in Chapter 1. Near the fusion line of the weld, a coarse grain zone (CGZ) exists. External forces such as magnetic stirring can enhance the macro-segregation and reduce the solidification rate through higher flow velocity of the molten Al [26]. This causes the CGZ to be split into two regions: CGZ-I and CGZ-II. A wider columnar region with coarser grains and a lower hardness is observed in CGZ-I, while the grains are finer and the hardness is higher in CGZ-II. This is due to pushing of the solute elements from CGZ-I into CGZ-II. The pushing of elements into the center of the melt pool further increases the constitutional super cooling, which, along with a lower temperature gradient, G, due to stirring, enhances the formation of finer equiaxed grains in the center of the FZ [26]. Figure 4.11a and b shows Electron Back Scatter Diffraction (EBSD) images of these weld zones without stirring and with stirring, respectively. CGZ-I is the preferred site for failure during mechanical testing due to its low hardness. Electrode geometry can also affect the morphology through providing path for heat

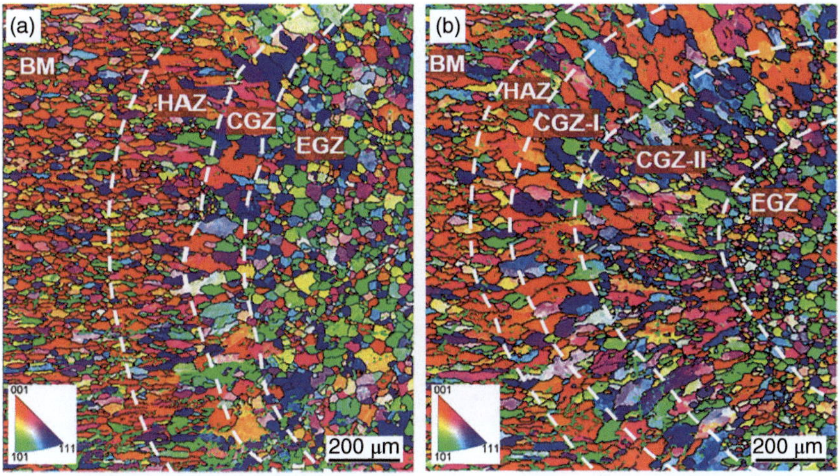

Figure 4.11 EBSD image of the weld zone of an Al–Zn–Mg alloy welded by (a) common resistance spot welding and (b) electromagnetically stirred resistance spot welding. Source: Ref. [26]/with permission of Elsevier.

sink and determination of the current density [27]. When the cooling effect of electrode is low, *G* becomes lower, which leads to a reduction in the region with coarse columnar grains [27].

The effect of current on the grain morphology in the nugget zone has multiple aspects. On the one hand, the higher current produces more heat, which may seem to lower the G and stimulates the columnar grains formation. But on the other, it softens the Al, which helps the electrode to indent more into Al, decreasing the distance between the water-cooled electrode and the weld center. This intensifies the water-cooling effect, helping to increase the G and hence promote the formation of finer and more equiaxed grains [28].

Besides the precipitate and grain size and morphology, the texture evolution in the FZ can also affect the mechanical properties of the FZ through the variation in the Taylor factor. A lower Taylor factor is indicative of a higher propensity for slip and a higher elongation [29]. Figure 4.12 shows how the crystal orientation in the FZ of an AA6061 weld has changed. The preferred orientation of slip planes ([111] planes) causes an increase in the Taylor factor in the FZ, meaning that the resistance to slip along these planes has become higher.

4.2.2 Thermo-Mechanically Affected Zone (TMAZ)

TMAZ is usually observed in solid-state welding processes such as FSW, which is partially affected by both heat and deformation. A similar region is observed during RSW of Al alloys, which is present between the base metal and nugget zone. This region is subdivided into three distinct regions as shown in Figure 4.13a.

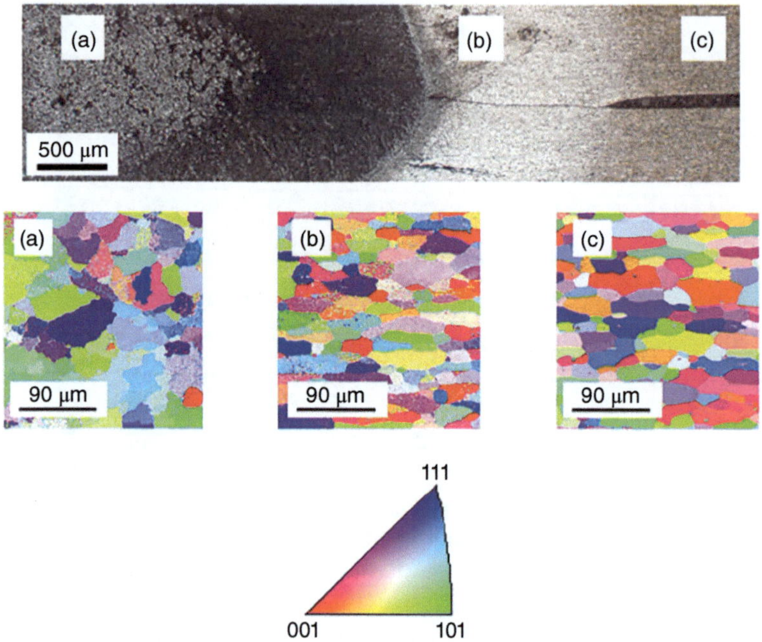

Figure 4.12 Inverse pole figures showing the crystal orientation in (a) FZ, (b) HAZ, and (c) base metal. The reference direction is the rolling direction. Source: Ref [29]/with permission of Elsevier.

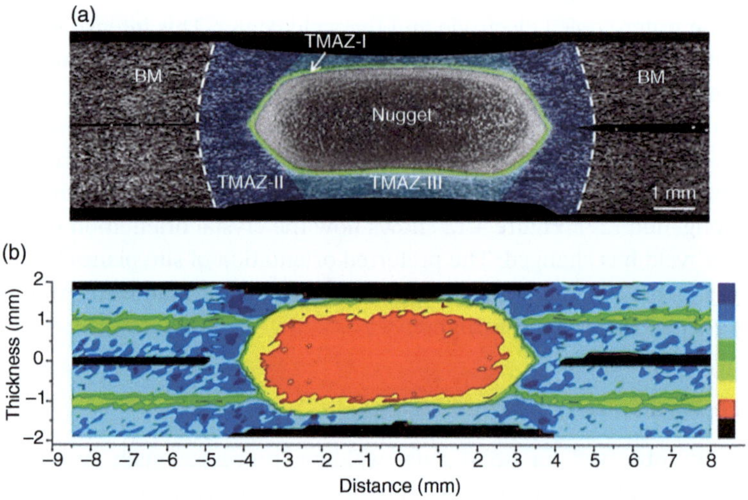

Figure 4.13 (a) Cross section of the welded sample of AA611, showing distinct regions of TMAZ. (b) Hardness map of this cross section. Source: Ref. [30]/with permission of Elsevier.

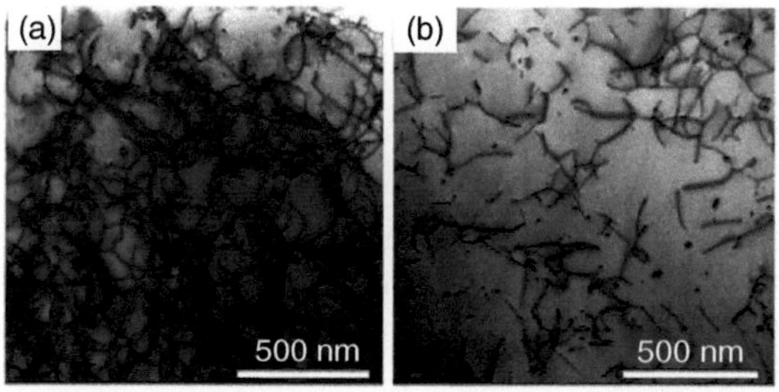

Figure 4.14 TEM image of regions (a) TMAZ-II and (b) the base material. Source: Ref. [30]/with permission of Elsevier.

TMAZ-I is located at the nugget boundary wherein eutectic constituents are melted along the grain boundaries and resolidified after welding, leaving a dendritic microstructure. The PMZ is in this region where the temperature exceeds the solidus temperature but is below the liquidus temperature. Large recrystallized grains are seen in this region [27]. TMAZ-II or hardened TMAZ is influenced by deformation and to a lower extent by the heat of electrode. This region is hardened by both deformation and precipitation. TMAZ-III is more affected by the heat of the electrode and to a lower extent by deformation induced by the electrode, showing a lower hardness than TMAZ-II and similar to the base material (Figure 4.13b) [30]. In fact, the strain hardening in TMAZ-II caused by the force of the electrodes causes the dislocation density to be raised compared with the base material. Due to insufficient time of heating resulting from a high cooling rate during RSW, no recovery can occur (compare the dislocation densities in regions TMAZ-II and base material shown in Figure 4.14a and b, respectively) [30]. While in TMAZ-III, due to higher temperature, recovery causes the hardness to drop. Al alloys that contain multiple elements and have precipitates show the PMZ during RSW as they have a temperature range of solidification in contrast to commercially pure Al (1xxx series) which have a fixed melting temperature [31].

4.3 Defects During RSW of Al Alloys

The defects are detrimental to the mechanical properties, especially fatigue behavior, when they are located in the periphery of the nugget zone. Usually, the large nugget diameter can compensate for the defect in terms of mechanical performance [32]. The cracks in the nugget zone are observed due to solidification shrinkage and high thermal expansion coefficient of Al, which are exacerbated by increasing the welding time [25]. The spatters may also be seen due to the higher expansion rate of the melt with respect to the surrounding plastic solid ring. This makes the liquid to break through the bondage of the plastic ring under the pressure of the electrode [22]. Solidification begins from the periphery of the melt, close to the HAZ, and continues to the center of the nugget. The pores form in the center of the nugget due to the lack of liquid to compensate the solidification shrinkage. There is a critical welding

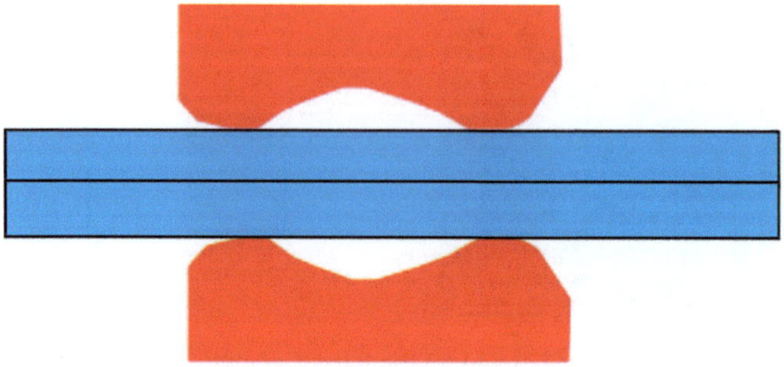

Figure 4.15 Schematic of an electrode with an outer ring and an inner convexity to inhibit splash formation.

current beyond which the expulsion occurs. The presence of the oxide layer on the contact surfaces of two sheets lowers this critical current, limiting the nugget size, which can be produced without expulsion [13].

The splash can be controlled by using an electrode with an outer ring and an inner convex shape, which directs the molten Al inward rather than outward due to the force exerted by the electrode's ring at the beginning of welding. The subsequent deformation of the Al in contact with the electrode's inner convex surface further extends the molten metal inward, forming a complete nugget (Figure 4.15) [22].

SC is another type of defect that appears when its conditions are met. SC in RSW of Al takes place under more complicated conditions than arc welding, as the heating and cooling occur in a much shorter time in RSW. Moreover, a steeper thermal gradient exists in RSW. Also, no filler can be used in this process to control the chemical composition of the weld nugget outside the hot cracking sensitivity region. RSW is a transient condition while arc welding of seams is a steady-state condition. Hot cracking can occur in the nugget zone or in the PMZ. When the solid fraction in the base metal is higher than the FZ at the final stages of cooling, hot cracking occurs in the PMZ [33]. The solidification cracks in the PMZ of RSW of Al are observable in the cross sections taken in longitudinal directions (Figure 4.16) [34].

The schematic of the thermal cycles in various regions of a spot weld is presented in Figure 4.17. These thermal cycles are matched with the phase diagram. PMZ is a region where the temperature lies between the solidus and liquidus temperatures. The grain boundaries in this region melt due to existence of low-melting-temperature constituents. The cracks made in this way may be filled by the melt from the FZ, before the current is shut off. As these cracks are outside the region where the pressure of the electrodes exists, they suffer local tensile

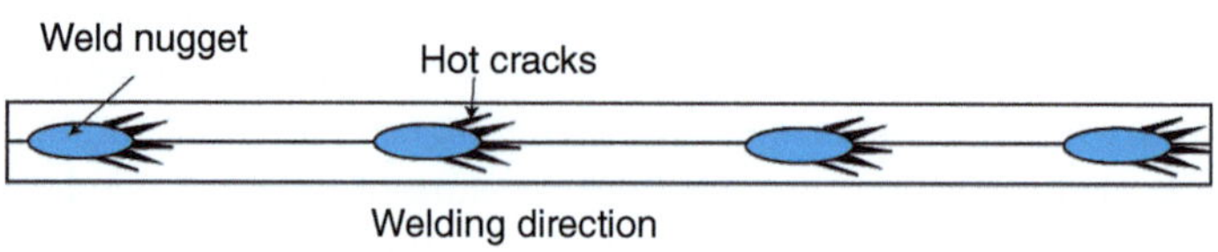

Figure 4.16 Schematic of hot cracks seen in the cross sections in the longitudinal direction.

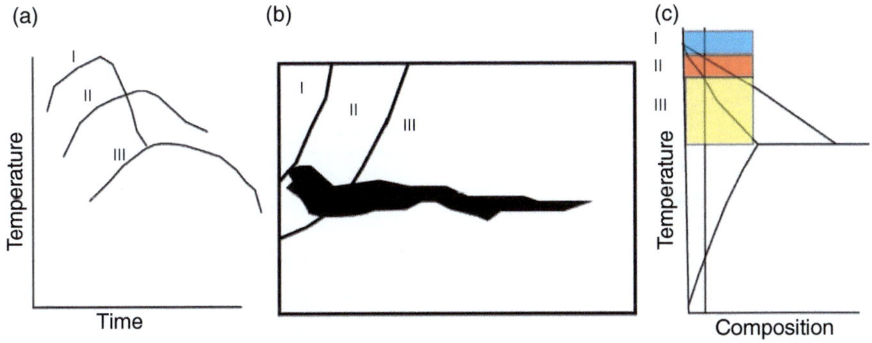

Figure 4.17 Schematic of (a) temperature–time history, (b) hot crack initiation in region II in a spot weld made on Al, matched with (c) phase diagram. (I is nugget zone, II is PMZ, III is base metal).

stresses and strains, making them more likely to propagate. To suppress the cracks propagation, a sufficient constraint around the weld spot is needed to apply pressure [35]. Between the PMZ and the base material where the temperature is below the solidus temperature, some liquation may also occur, which is different from the liquation in the PMZ. In this region, the high heat rate during RSW cannot dissolve the precipitates in the matrix. So, the precipitates are left in an un-equilibrium state above the eutectic temperature. This causes very localized liquation around these precipitates, though the temperature is below the solidus [34]. No report was observed for SC in the FZ of Al alloys during RSW. This is most probably due to two main reasons: First, the last place to solidify in the FZ is the center, which solidifies in the equiaxed mode, that is, a microstructure that is less susceptible to SC. Second, the force of electrodes closes the cracks induced by the SC, if any. In case there is any crack in the FZ, the initiation sites are from the PMZ, where liquation cracks form and propagate into the FZ [36].

The weld nugget in RSW joints is usually circumferentially symmetric. Some factors such as the stiffness of the electrode arms may cause this symmetry to be distorted. For example, the C-gun arms (Figure 4.18a) used in the automotive industry are deflected during the process, causing a nonuniform distribution of force and current at the joint interface in the inner side and outer side, as shown in Figure 4.18b [37].

4.4 RSW of Dissimilar Aluminum Alloys

The shape of the nugget in RSW of similar materials with the same thickness and by using the same electrodes on the two surfaces is a symmetric lens shape. The diameter of this lens is equal to the diameter of the electrodes. In the case of dissimilar thicknesses, the heat balance is distorted and the sheets are heated up unequally, which leads to the formation of an asymmetrical nugget. To prevent this asymmetry, the diameter of the electrodes should be different (Figure 4.19). The relative size

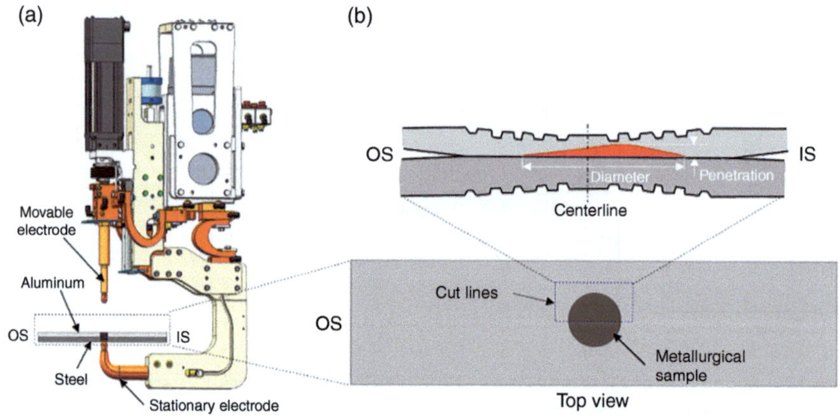

Figure 4.18 (a) C-gun arms of a RSW machine. (b) Asymmetry of the nugget on the inner side (IS) and outer side (OS) of the joint nugget. Source: Ref. [37].

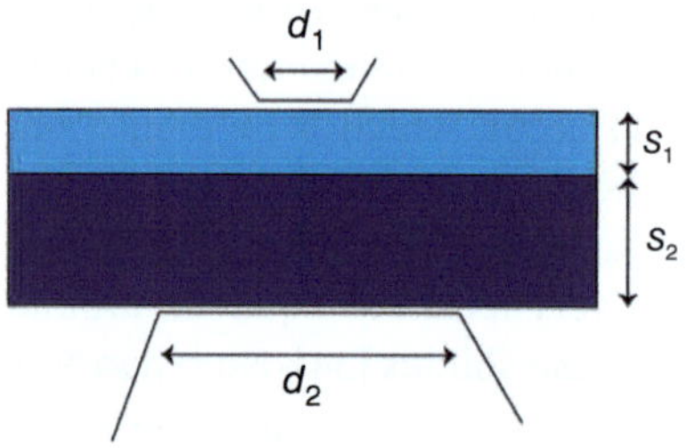

Figure 4.19 Diameter of the electrodes for joining two dissimilar thicknesses.

of the electrodes is obtained from [38]:

$$d_2 = \frac{d_1 s_2}{s_1} \tag{4.5}$$

where d_1 and d_2 are the diameters of the electrodes in contact with the sheets with the thicknesses of s_1 and s_2.

During RSW of dissimilar steels, such as carbon steel to stainless steel, the heat conductivity plays an important role in nugget formation. The electrical and thermal conductivity of stainless steel up to 800 °C is lower than carbon steel, leading to higher heat generation on the side of stainless steel, and therefore, an asymmetric weld nugget appears (Figure 4.20). To obtain a nugget with the best asymmetry, the welding parameters such as the diameter of the electrodes in contact with each material need to be adjusted [39].

The same asymmetry is observed during RSW of two dissimilar Al alloys, for example a wrought AW-AlSi1MgMn and a cast AC-AlSi7MgMn (Figure 4.21). This is due to a lower melting range and a higher electrical resistance of the cast alloy. Also, solidification cracks exist only in the HAZ of wrought Al alloy [5]. Similar behavior is observed during RSW of Al alloys with dissimilar thicknesses wherein the sheet with the smaller thickness melts more than the thicker one.

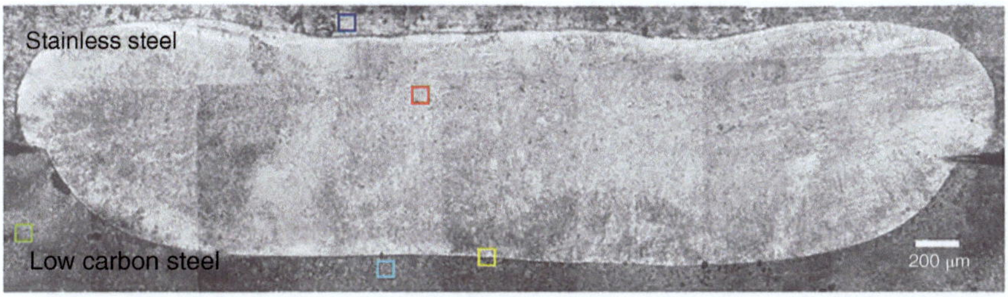

Figure 4.20 Asymmetric weld nugget during RSW of stainless steel to carbon steel. Source: Ref. [39]/with permission of Elsevier.

The Peltier effect during RSW of dissimilar materials is not the predominant factor in the asymmetry distortion of the FZ. Instead, the bulk resistivity of the two Al alloys plays an important role in the shape of the nugget and its asymmetry. For instance, during RSW of AA5052 and AA 6014, due to higher thermal conductivity of AA6014 and its higher cooling effect, a smaller nugget forms in AA6014. Also, the bigger nugget in AA5052 causes a smaller distance between the nugget and the electrode and hence a higher temperature is seen in the electrode in contact with AA5052. This consequently causes a faster degradation of this electrode (Figure 4.22a) [40]. The cracks due to liquation cracking are observed in the HAZ of AA5052 due to its susceptibility to SC (Figure 4.22b–e) [40].

The usage of a thin dissimilar interlayer with a lower melting point during RSW of Al alloys is beneficial in lowering the required current to obtain a determined nugget size. This is because of the higher heat resistance due to existence of two interfaces as well as the lower melting temperature of the insert layer [41]. An example is usage of a thin Al—Si foil between two Al—Mg—Si alloys by which the composition and microstructure of the nugget are improved [41].

4.5 RSW of Al to Dissimilar Metals

Direct welding of Al to other dissimilar metals is associated with IMC formation, which degrades the mechanical properties and performance of the joint [42]. The type and morphology of the IMCs (grain size) determine how brittle the

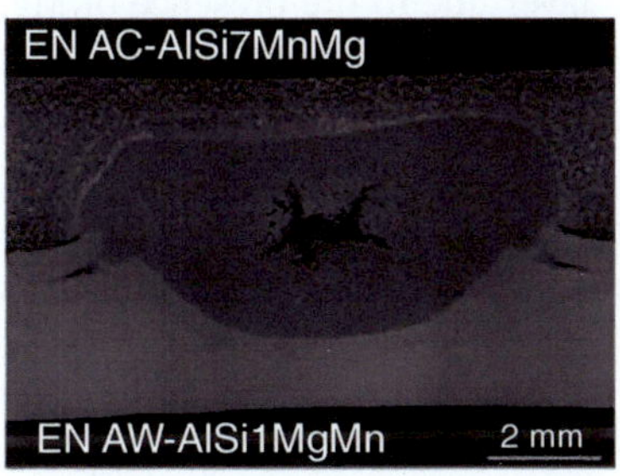

Figure 4.21 Macro-image of spot weld between two dissimilar Al alloys, a wrought AW-AlSi1MgMn and a cast AC-AlSi7MgMn. Source: Ref. [5]/Sage Publications.

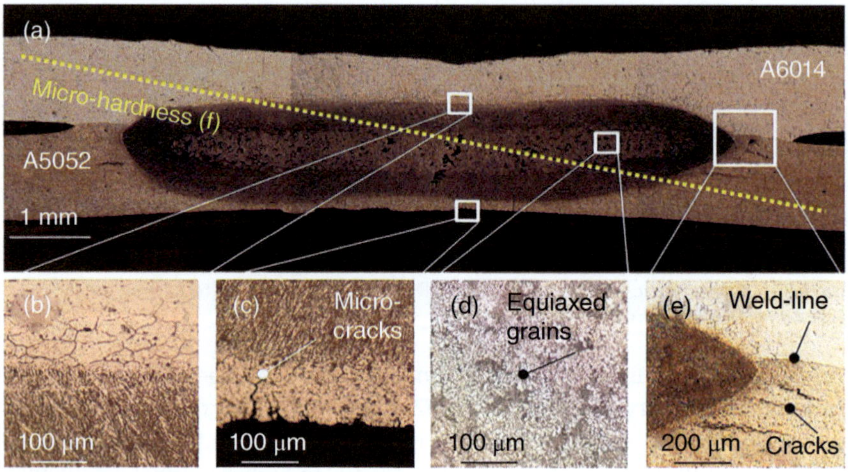

Figure 4.22 (a) Cross section of the nugget in a joint between AA5052 and AA6014 made by resistance spot welding. (b–e) Higher magnifications of the various locations showing cracks in the HAZ of AA5052 due to liquation cracking. Source: Ref. [40]/Taylo & Francis/Sage Publications.

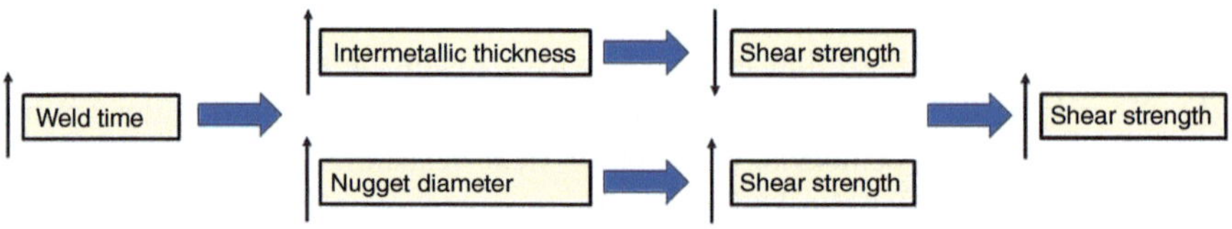

Figure 4.23 Effect of welding time on the nugget diameter and IMC thickness, and shear strength of the lap joint between Al and steel.

interface is [43]. Several factors affect the IMCs, with the compositions of the parent materials being among the most important. For instance, the IMC formation varies depending on whether Al is joined to carbon steel or to a higher-strength, low-alloy steel [44]. Also, the Al type can affect the IMCs when welded to the same steel. AA6xxx produces a thin IMC layer due to the presence of Si, which inhibits the diffusion, while AA5xxx produces a thicker IMC layer due to the presence of Mg, which promotes the growth (Figure 4.24c, d) [45]. However, compared to other joint designs such as butt joints, IMCs are not a critical factor in degrading joint strength in RSW of dissimilar metals performed in a lap configuration. This is because the loading condition in a lap configuration is primarily shear. As a result, joint strength can be improved by increasing the nugget diameter, even though the thickness of the IMC layer may increase due to higher heat input. In other words, joint strength is not compromised by the brittleness of thick IMCs; instead, it is enhanced by a larger nugget diameter (Figure 4.23) [46]. The common IMCs in Al/steel joints made by RSW are FeAl, Fe_2Al_5, and Fe_4Al_{13}. FeAl is an iron-rich IMC that has a lower hardness and higher ductility [47].

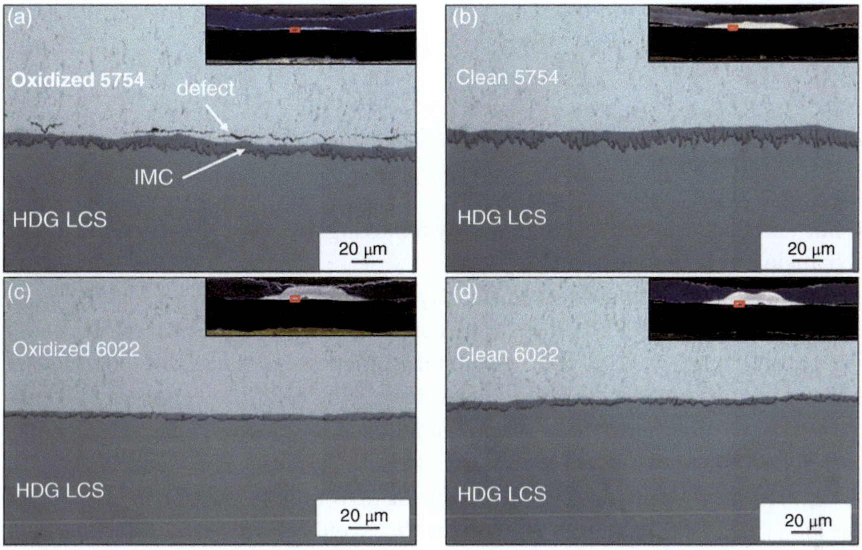

Figure 4.24 Interface between Al and steel jointed by resistance spot welding. (a) AA5754 with oxide layer. (b) AA5754 with cleaned surface. (c and d) AA6022 with and without oxide layer. Source: Ref. [45]/© American Welding Journal.

The oxide layer on the surface of Al affects the joint quality when welded to steel, like when welded to Al. The oxide layer at the faying surface is not melted during RSW and will be present in the solidified nugget zone of Al close to the interface (compare Figure 4.24a, b). This oxide layer affects the strength and the crack propagation path in the joint. The oxide layer between electrode and Al affects the resistance and hence the expulsion effect and nugget size. When AA5xxx is welded to steel, due to it thicker oxide layer, higher resistance is made which causes expulsion and reduced nugget size compared to AA6xxx [45]. Figure 4.24 clearly shows why removing the oxide layer is more critical in some Al alloys such as 5xxx alloys.

IMCs limit the application of RSW for joining of Al to other structural materials such as steel in body structure of the cars where a cyclic load is present [48]. The role of IMC in the degradation of the joint strength made by RSW is less critical than other techniques that are mainly performed in butt configuration. This is because the IMCs are stronger in shear than in normal loading conditions [49]. The periphery of the nugget zone acts as the initiation site for crack propagation due to the notch effect in this region [50]. Increasing the angle of the notch root by a proper strategy can mitigate its detrimental effect on the mechanical properties of dissimilar joints. Usage of a shim to introduce a gap between the sheets increases this notch angle, necessitating a large deformation of Al during RSW (Figure 4.25) [51]. Though this method raises the expulsion effect in the periphery of the joint, it increases the tensile shear strength due to reduction in the stress concentration. The expulsion also cleans the oxide layer at the periphery of the nugget, which has a positive effect on the joint strength. So, the overall effect of the gap between the dissimilar metals on the joint quality is competitively determined by the negative effect from the expulsion from

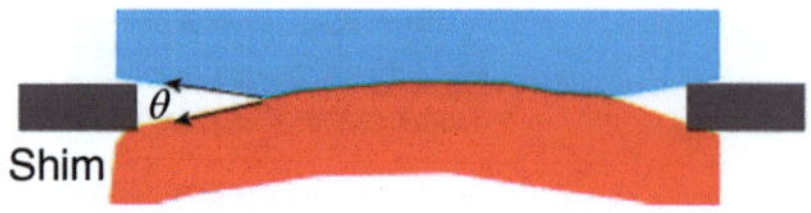

Figure 4.25 Application of a shim to increase the angle of the notch during resistance spot welding of Al to steel.

one side and the positive effect obtained by removing the oxide layer and raising the notch angle from the other side [51].

The key factor in joining of Al to steel by fusion welding processes is to control the thickness of the IMC. To do so in the conventional RSW, the heat input should be kept as much as low. This approach, on the other side, reduces the spot size and hence the joint strength is declined. By increasing the heat input (time or current), though the spot size increases, a massive amount of IMC forms in the nugget zone. The form of IMCs depends on the materials and heat input. For example, Al—Cu IMCs appear as an eutectic of $CuAl_2$—Al during RSW of Al to Cu, causing brittleness of the joint [52]. Weld-brazing is a technique that avoids melting of steel during RSW of Al to steel to control the IMC thickness, as the kinetics of the growth of IMCs is higher when both Al and steel are in the liquid state. The same holds for joining of Al to other materials such as Cu where the melting of copper accelerates the IMC formation and growth [53]. Usually, in conventional RSW of Al to steel, steel melts away from the Al/steel interface, which helps to control the detrimental IMC growth (Figure 4.26) [47]. In other fusion welding processes where a filler material can be added, such as TIG or laser welding, the addition of other elements controls both the type and the thickness of the IMC layer [54]. The most common filler materials for weld-brazing of Al to steel are Al—Si and Zn—Al filler wires [55, 56]. The addition of filler material during RSW is challenging and therefore other strategies such as coating is used to control the chemical composition. Using Zn-coated steel made by hot galvanizing is a method to do so by which the Al—Fe IMC compounds contain Zn such as $Fe_2Al_5(Zn)$ and $FeAl_3(Zn)$ [57].

To minimize IMC formation at the interface, one approach is to use an insert between Al and steel. This insert is first joined to the steel using a solid-state welding method, such as ultrasonic welding, to limit IMC formation. Then, the cladded steel is joined to Al by RSW. A larger, flat-surfaced electrode is applied to the steel to distribute current over a wider area, reducing heat build-up in the steel. This method

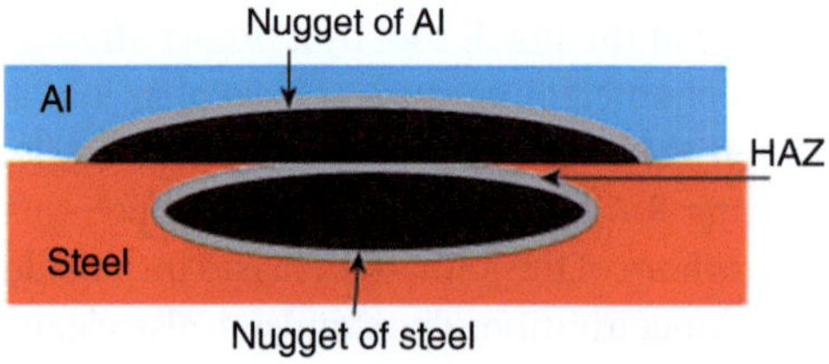

Figure 4.26 Schematic of cross section of resistance spot welding between Al and steel. Melting in steel occurs away from the interface.

minimizes expulsion and forms a brazed joint rather than a welded one. In other words, the steel remains solid while the molten Al spreads across its surface [58].

In another approach, a thin pure Al sheet is roll-bonded to the surface of the steel, making an intermediate layer for the subsequent RSW of Al to steel. In this method, no melting of steel occurs, helping to control the thickness of the IMCs and hence increase the load-bearing capacity of the dissimilar joint [28]. The interlayer can also be cladded on the surface of Al. A technique used to deposit an interlayer on the Al is electro-spark deposition. In this technique, a short-duration arc is formed between a rod and the workpiece by discharging a capacitor. The droplets of the rod are transferred from the rod to the workpiece, where due to short pulse duration the heat input is very low and the cooling rate is very high [59]. Low-pressure cold spraying is also used to make an interlayer before RSW, having the advantage of using powders containing both metals and ceramics or a mixture of them [60]. Making a metallic bump by CMT technology on the surface of Al is another method to make an interlayer between Al and steel for RSW [61]. This metallic bump is squeezed out of the interface during RSW of Al to steel, helping to crush the oxide layer and squeeze it out. This bump also confines the electricity path, causing higher heat generation and larger nugget diameter [62]. Other techniques such as inserting graphene layer between steel and Al hinder the solubility of Al into steel and at the same time enhance the strength by making a nanocomposite containing nano-platelets of graphene [49, 63].

The key point in using an interlayer for RSW of Al to steel is to limit the heat input during deposition or joining the interlayer to the surface of steel, so a very thin IMC is formed at the interface of steel and the interlayer. In subsequent RSW of Al to this steel, the heat is generated at the interface of Al/interlayer, which helps to prevent any direct reaction between Al and steel in the liquid state. Furthermore, the heat that is generated at the interface of the interlayer and Al is spent to melt two Al alloys in contrast to direct RSW of Al to steel, where the heat is generated at the interface of Al and steel and accelerates the IMC formation.

Enhancing the mechanical locking between two dissimilar materials by using innovative electrodes is another way to compensate for the weak joint caused by IMCs (Figure 4.27) [50]. Further improvement of this joint was attained by doing laser welding in the periphery of the nugget zone to overcome the weak point in the periphery.

In cases where steel is very hard, it is difficult and nonpractical to weld an Al insert to steel. Here, another approach can be used by which the interlayer is welded to Al rather than steel. In this case, the material used as the insert is not Al, but another steel such as stainless steel. This insert can be welded to Al by ultrasonic welding, and then this combination is welded to the coated press-hardened boron steel by RSW. During the RSW process, Al is fused on the surface of the insert making a

Figure 4.27 Schematic of joint showing the bulge of steel into Al, making a mechanical interlocking.

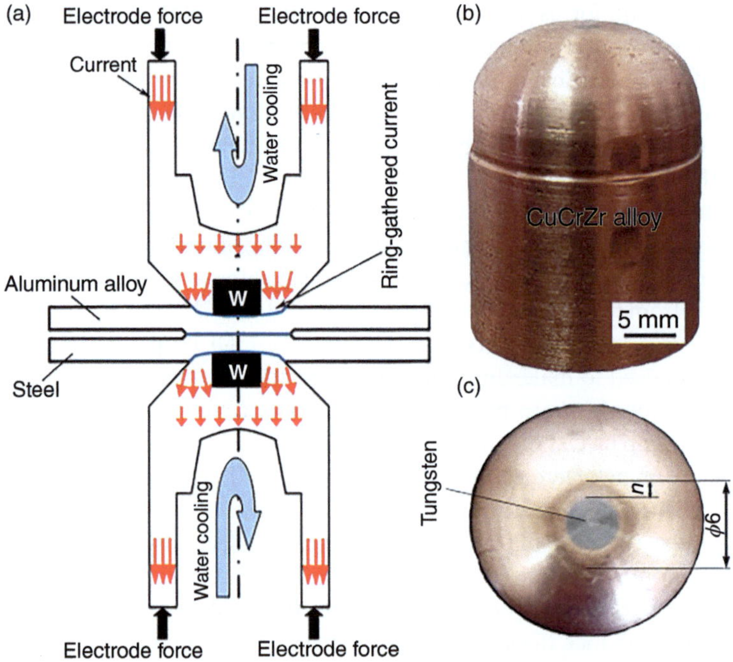

Figure 4.28 Usage of a composite electrode with a tungsten rod embedded in the center of the electrode. (a) Schematic view of welding process. (b) Side view of the electrode. (c) Bottom view of the electrode. Source: Ref. [65]/with permission of Elsevier.

weld-brazed joint with IMCs at the interface, which determines the overlap joint strength. The steel and the insert previously joined by ultrasonic welding fuse at their interface as well, making the coating on the steel to disappear [64].

Another approach for RSW of Al to steel is modification of the electrode. The electrode is a composite electrode made of copper, the center of which is embedded with a tungsten rod (Figure 4.28). Tungsten in the center of the electrode, due to its lower electrical conductivity with respect to the copper, produces less heat in the center, and hence the IMCs in the center of the weld interface become thinner [65].

Apart from mechanical properties of the joints between Al and other materials, the corrosion resistance is another important factor, as these joints are prone to galvanic corrosion. Using adhesive bonding along with RSW to join Al to steel can alleviate this problem by eliminating direct contact of these two metals [66].

4.6 Modifications to RSW for Welding Al Alloys

Due to complexities resulting from the presence of oxide layer for the contact resistance, several modifications have been made to the process. Resistance spot clinching (RSC) is the one that combines both welding and mechanical clinching and at the same time eliminates the limitations raised from the oxide layer. The electrodes in this process are shaped like molds to deform the sheets and make

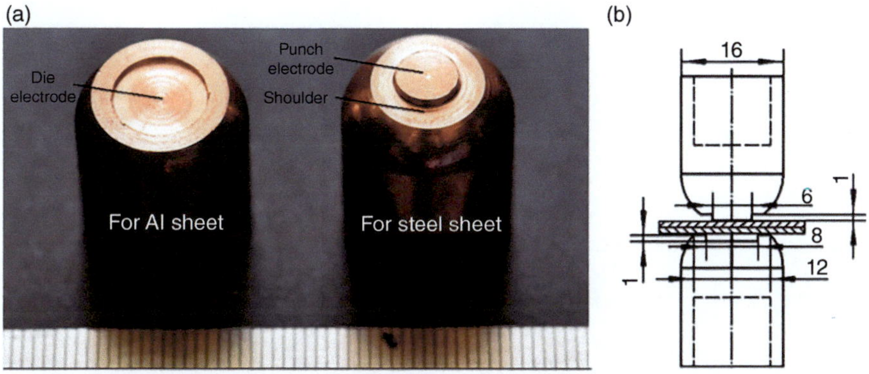

Figure 4.29 (a) Electrodes used for resistance spot clinching. (b) Schematic of the process (dimensions in mm). Source: Ref. [69]/Sage Publications.

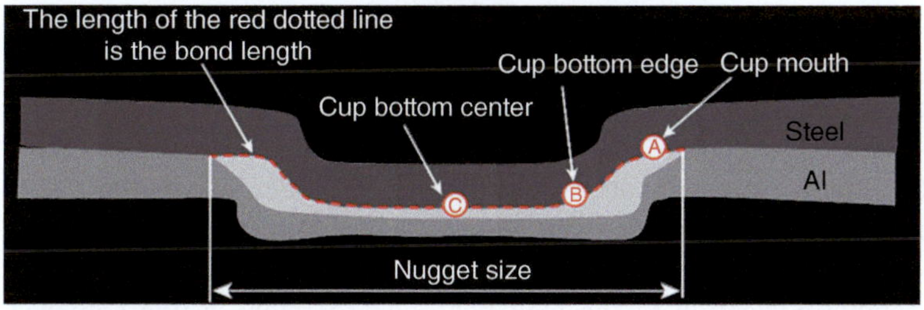

Figure 4.30 Schematic of a resistance spot clinching joint between Al and steel. Source: Ref. [69].

a mechanical locking. Figure 4.29 shows the shape of the electrodes. Very thin tapes made of AISI 201 are placed between the modified electrodes and the Al sheets, whose interface with Al is the main source of heat generation [13]. Two cycles of currents are applied during this process [67]. The first current cycle is for softening the sheets in the solid state to facilitate the deformation and clinching. The second current cycle is applied to melt the Al sheets and make a weld by subsequent solidification during the cooling stage. RSC is less sensitive to the oxide layer, as the heat is mainly produced at the interface of the steel tape and Al, rather than the interface of two Al sheets. The schematic of the cross section of a resistance spot cinching is shown in Figure 4.30. In a conventional spot joint, the rotation of the joint during shear tensile test causes a decrease in the peak load, as the additional components of stress introduced by the rotation (σ) leads to a lower critical shear load (τ) according to von Mises criterion:

$$\frac{1}{3}(2\sigma^2 + 6\tau^2)^{0.5} = \frac{\sqrt{2}}{3}\sigma_f \tag{4.6}$$

where σ_f is the fracture stress.

By RSC, a higher stiffness of the joint inhibits rotation during the shear tensile test, and therefore, a higher shear stress is needed to plasticize the joint, leading to a higher peak load [68].

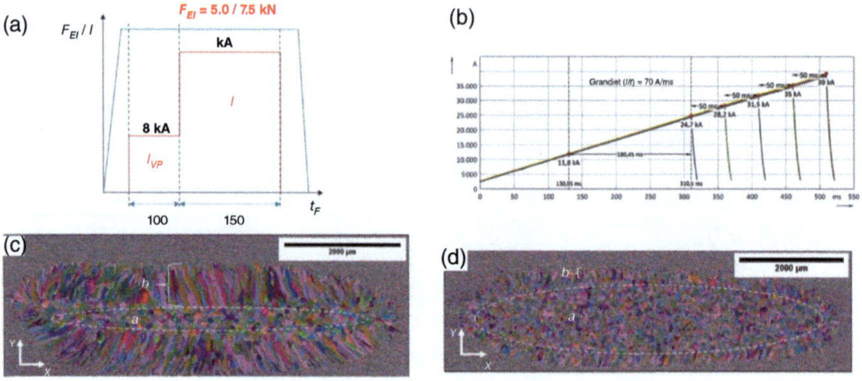

Figure 4.31 Welding cycles for (a) conventional welding according to VDA 238-401 and (b) modified upslope. Resulting microstructures made by (c) conventional cycle and (d) modified upslope cycle. Source: Ref. [71]/with permission of Elsevier.

Application of ultrasonic during RW facilitates surface deformation at the electrode/Al interface, helping to break the oxide layers and lower the interface resistance [70]. This is caused by acoustic-plastic effect, which makes the deformation of surface asperities much easier due to lower flow stress of material by ultrasound application. The cavitation induced by ultrasonic vibration promotes heterogenous nucleation in the melt pool, and therefore, a fine equiaxed microstructure is developed in the nugget. The acoustic streaming also helps to fill the holes caused by solidification shrinkage. The threshold ultrasonic intensity (I) to induce acoustic cavitation in Al melt is equal to 80 W/cm^2. The acoustic intensity is obtained from [70]:

$$I = 0.5\rho c(2\pi f A)^2 \tag{4.7}$$

where ρ is the density of the melt, c is the speed of sound in the melt, f is the frequency, and A is the vibration amplitude. Acoustic wave produces alternate low-pressure and high-pressure fields inside the melt. Cavitations are produced in the low-pressure region, which collapses during the high-pressure period and produce shock waves that break the columnar structure during solidification and provide nucleation sites for equiaxed grains [70].

RSW is usually used for 5xxx and 6xxx grades. Other grades such as 7xxx are prone to SC and are not suitable to be joined by any fusion welding process, including RSW. A modification made to the current cycle, named upslope, can lower the risk of SC by extending the fine-grained equiaxed zone. The cycles of conventional and upslope current are shown in Figure 4.31a and b, respectively. The resulting microstructures of the nugget zone are shown in Figure 4.31c and d, respectively. The reason for the formation of a dominant FQZ by modified upslope cycle is a reduced G/R ratio [71].

External magnetic field can alter the microstructure of the nugget zone favorably. The way the magnetic field is applied is depicted in Figure 4.32. A pair of repulsive magnets produce an external magnetic field (B_{ext}) that creates a magnetic force (F_{ext}) according to the Fleming's left-hand rule. This force is in radial direction. There

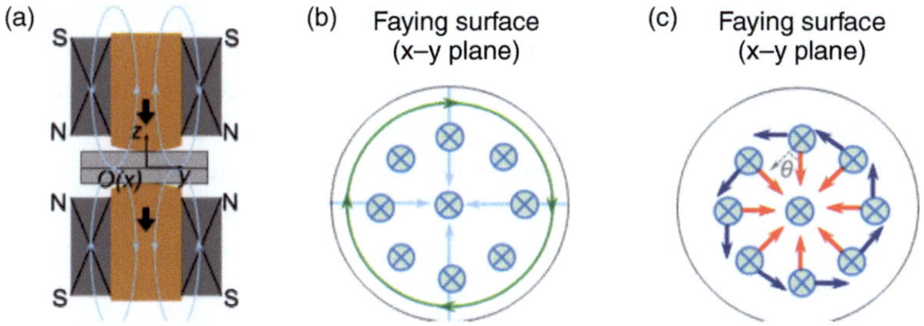

Figure 4.32 (a) Schematic of external magnetic field and the current path during RSW of Al. (b) Centripetal and radial forces. (c) Resulted force. Source: Ref. [72].

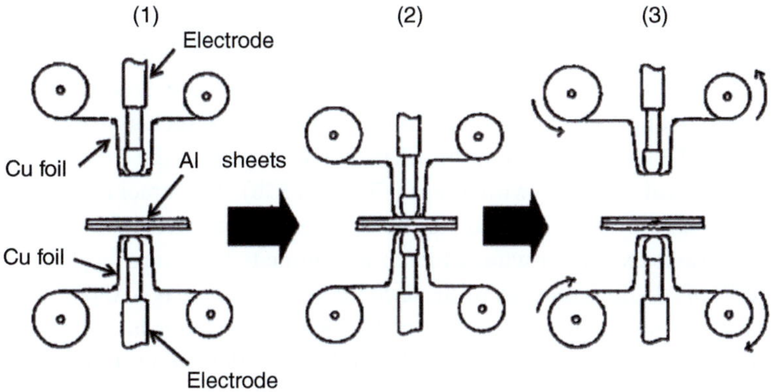

Figure 4.33 Using a Cu-foil as a shield for electrode protection during RSW of Al alloys. Source: Ref. [11].

is another magnetic field caused by the current passing through the electrodes according to the Ampere's right-hand screw rule. This induced magnetic field (B_{ind}) is in radial direction and creates an induced magnetic force (F_{ind}), which is centripetal. The direction of the resulting force of these two forces depends on the relative magnitudes of these forces, which determines the flow direction inside the melt pool [72].

The quality of the welds on Al when using the conventional procedure is not the same for all the spots, as the electrode degrades after each weld due to alloying effects. To solve this issue, a coated thin tape is used between the electrode and the workpiece. This tape eliminates the contamination effect and improves the electrode life, as the contact surface is always clean due to movement of this tape after each spot weld. This tape also improves the contact resistance and makes the spot to be reproducible (Figure 4.33) [73]. Also referred to as thermo-compensating tape, it can be made of a less expensive material such as steel to reduce the heat sinking effect of workpieces made of Al or copper in contact with the electrode [52].

The high resistance of the tape provides a higher temperature at the interface of the electrode and the workpiece, as shown in Figure 4.34. During dissimilar welding, such as Al to steel, the tape between the Al and electrode has a higher resistance than the one between steel and electrode due to higher conductivity of Al. The use of this tape extends the acceptable range for RSW of Al to steel [74]. It is recommended to use a tape made of austenitic stainless steel between Al and electrode, while a tape of carbon steel is better to be placed between steel and electrode. This is because stainless steel has higher resistance and lower thermal conductivity, making it more suitable on the Al side [46]. The usage of the tape also lowers the indentation of the electrode into Al, causing a higher distance between the FZ and the electrode, which further helps to increase the electrode life. In general, it is recommended to use different tapes with different electrical resistivity when welding Al to other materials. The tape in contact with Al needs to have a higher resistivity [75].

Al alloys such as 7xxx are prone to SC during RSW, as no filler material can be added to the weld to control the chemical composition. A combination of mechanical locking and metallurgical bonding named resistance plug welding can alleviate this problem in RSW of 7xxx Al alloy. Two holes are made on both sheets of 7xxx alloy to insert a filler rod of another Al alloy that has better weldability such as 5xxx alloy, as shown in Figure 4.35a. During RSW, this filler wire is fused into the 7xxx sheets, making a metallurgical bonding with them (Figure 4.35b). Hot cracking in the nugget zone is avoided by this method, as the composition of melt in resistance plug welding is close to 5xxx filler wire, which is not prone to hot cracking [76].

SC during RSW of Al alloys may occur in susceptible alloys such as AA6061. This is more probable during welding Al alloys to steel where the latter is not melted during RSW, and hence, there will be a lack of liquid to backflow and fill the solidification cracks. In these cases, the SC is observed parallel and close to the interface (Figure 4.36a). The application of ultrasonic resistance welding (URW) shown in Figure 4.36b helps to mitigate this by several mechanisms. The vibration increases the flowability and promotes backfilling. The cavitation effect breaks the dendrites and promotes equiaxed nucleation in front of the solidification, which lessens the SC (Figure 4.36c). The vibration also homogenizes the element distribution in the melt, which minimizes elements segregation and hence reduces SC susceptibility [77].

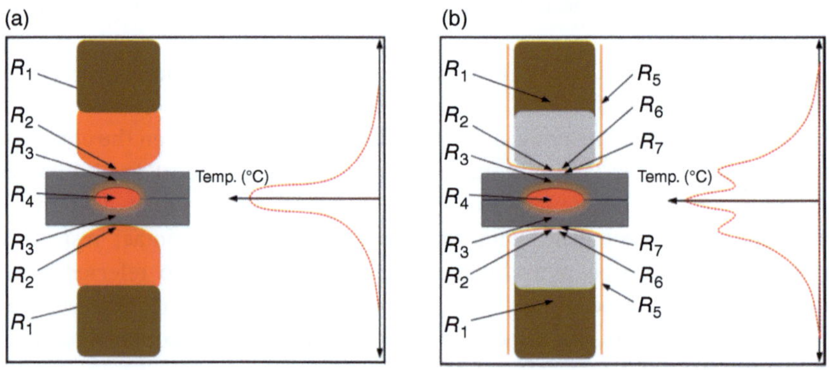

Figure 4.34 Temperature distribution during (a) conventional RSW and (b) RSW with a tape (*R* is electrical resistance). Source: Ref. [74].

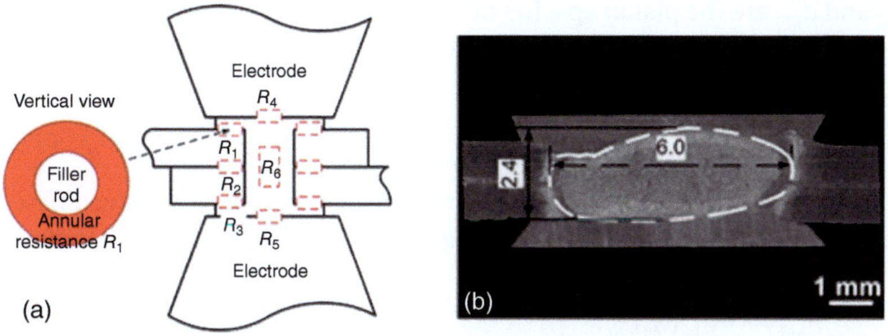

Figure 4.35 (a) Schematic of resistance plug welding of 7xxx Al alloy. (b) Weld cross section. Source: Ref. [76]/Taylor & Francis Group.

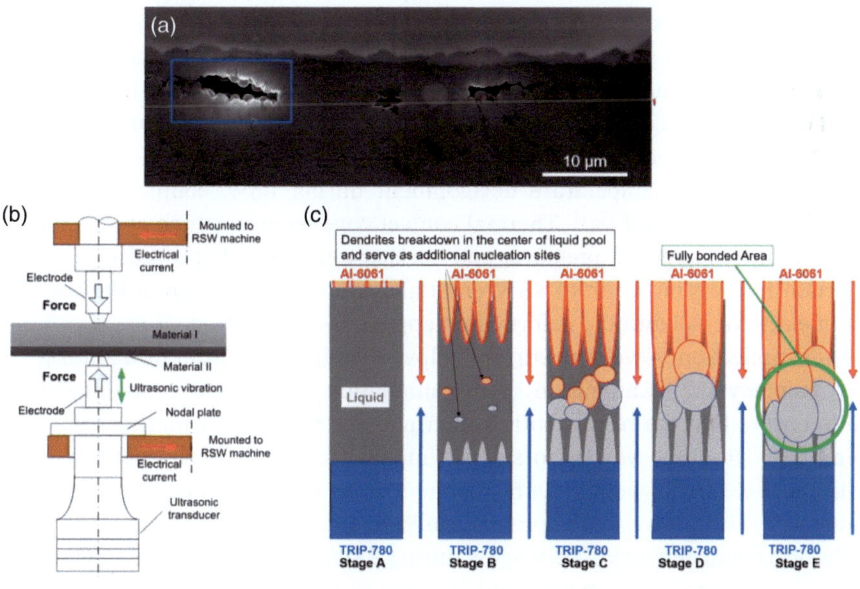

Figure 4.36 (a) Solidification crack close to Al–steel interface in Al joined by RSW. (b) Schematic of URW. (c) Process of break of dendrites in solidification front. Source: Ref. [77]/with permission of Elsevier.

4.7 Simulation of the Process

Residual stress during RSW takes place during the cooling stage where non-uniform heating along with the applied force of the electrodes contributes to it. The residual stresses developed during RSW of Al alloys are mainly in-plane, meaning that the residual stress in the thickness direction is low [78]. For measuring the residual stress, a specimen free of residual stress is needed to calculate the developed residual strain (ε_{hkl}) according to the following eq. [78]:

$$\varepsilon_{hkl} = \frac{d_{hkl} - d_{hkl}^0}{d_{hkl}^0} \quad (4.8)$$

where d_{hkl} and d^0_{hkl} are the planar spacing of hkl plane in both the welded specimen and free-strain specimen, respectively. The planar space is calculated according to the Bragg's law:

$$\lambda = 2d_{hkl} \sin \theta_{hkl} \tag{4.9}$$

where λ is the wavelength of the incident ray (neutron or X-ray) and θ_{hkl} is the diffraction angle from hkl plane. Then, Hook's law is used to calculate the residual stress (σ_{ii}) from the residual strain:

$$\sigma_{ii} = \frac{\varepsilon_{ij} E_{hkl}}{1+\nu} + \frac{\varepsilon_{kk} \nu \delta_{ij} E_{hkl}}{(1+\nu)(1-2\nu)} \tag{4.10}$$

where E_{hkl} is the elastic constant specific to *hkl* plane, ν is the Poisson's ratio, ε_{kk} is the summation of three strains in the main principal directions, and δ_{ij} is the Kronecker delta. Neutron diffraction and X-ray diffraction are two non-destrcutive methods to measure the residual stress. But these methods are very expensive and need a lot of time. Simulation methods based on FE are less costly and can give the distribution of residual stress in the entire region of the weld. SORPAS is a commercial FE software for simulation and optimization of RSW [79].

For prediction of temperature development during RSW, coupled electrical-thermal models are used [80]. Thermal contact conductance, which considers the influence of temperature, pressure, and surface roughness, is proved to be very important in the simulation of RSW of Al alloys [81]. The change in the contact area of electrode/Al with the deformation applied by the electrode force at various times of application of welding current helps to predict the nugget formation with more accuracy [82]. Calibration of the simulation parameters is an important step to ensure that the results both in thermal-electrical and in mechanical steps are reliable. The best way to do so is to match the nugget zone obtained from the thermal-electrical module with the one seen in the cross section of the weld (Figure 4.37) [12]. However, the presence of PMZ around the nugget alters the thermophysical characteristics of this region during RSW, which causes a discrepancy of the nugget size obtained from the simulation and experiment. In fact, due to the presence of liquid, this region acts as a heat barrier [31]. The solid and liquid fractions (f_s and f_l) in the PMZ can be approximated based on the Scheil–Gulliver model [31]:

$$f_s = 1 - \left(\frac{T_m - T}{T_m - T_L}\right)^{\frac{1}{k-1}} \tag{4.11}$$

The thermal properties (heat conductivity, thermal conductivity, and enthalpy) in the PMZ can be calculated by interpolating, knowing the solid fraction. Taking this into account, a more accurate shape of the nugget zone is predicted when welding two dissimilar Al alloys with different PMZ characteristics [31]. It should be mentioned that this approach has its own errors, as the formula pertains to the equilibrium state, while during RSW the heating and cooling rate are very high and the condition is non-equilibrium. Moreover, liquation in the PMZ occurs locally and not in the entire region, which adds another error to the formula, which assumes that the PMZ is uniform.

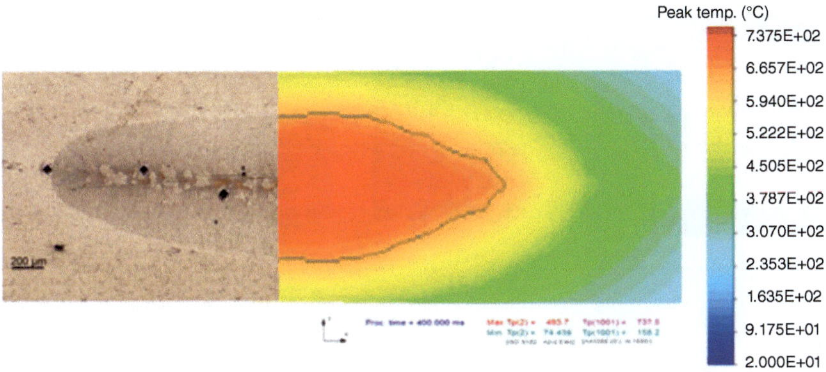

Figure 4.37 Cross section of the nugget and the predicted nugget shape from the simulation. Source: Ref. [12]/Taylor & Francis/CC BY 4.0.

During RSW of dissimilar Al alloys, another factor related to the FZ makes the simulation results to deviate from the experiments. As the FZ is asymmetric, the percentage of melting of each alloy imparted to the FZ is not equal. So, the chemical composition and therefore the material properties of the FZ need to be calculated correctly by knowing the percentage of each alloy. Each property in the FZ (P_{FZ}) is calculated from [31]:

$$P_{FZ} = xP_{Alloy1} + (1 - x)P_{Alloy2} \tag{4.12}$$

where x is the proportion of alloy 1 in the FZ. A calibration process is needed to attain the proper x value by setting an initial value to x (usually 0.5) in the simulation and matching the result with a known experiment [31].

Mechanical model by using the temperature history in thermal-electrical model provides the data for residual stress. The distribution of residual stress in the cross section of a weld on AA6061-T6 made by RSW shows that the highest residual stress is observed in the center of the FZ, decreasing toward the periphery of the nugget [83, 84].

The evolution of various microstructures in the weld zone of Al alloys joined by RSW can be understood by simulation results where thermal history is obtained. The occurrence of PMZ, precipitate dissolution, and recovery are the phenomena that can be estimated using the maximum temperature map and temperature–time history, as shown in two examples in Figure 4.38.

The simulation is a fast and cost-effective way to optimize the parameters of RSW. First, the model needs to be calibrated with some experiments, and then it can be used to obtain the spot diameter at various welding times and currents determined by design of experiments made at various thicknesses [85]. The simulation approach is more significant when the parameters of the process are numerous. For instance, when joining dissimilar materials such as Al to steel, the shape of the electrode and their size on two sides are not similar and a variety of electrodes combinations exist, which are very costly and time-consuming to investigate by experiments [86].

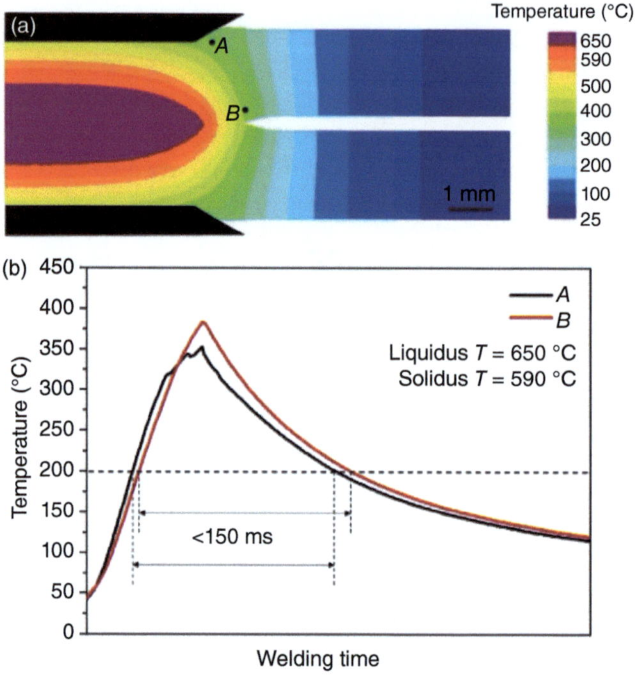

Figure 4.38 Maximum temperature map and temperature–time history of two points (*a* and *b*) during RSW of an AA6111-T6 alloy. Source: Ref. [30].

4.8 Evaluation of the Joints

The strength of the lap joints made by RSW are evaluated in the tensile shear test, as depicted in Figure 4.39. Considering the Mises criterion, the shear stress to cause plasticization and failure is $\tau_f = 0.577\sigma_f$. Hence, the load-bearing capacity of these joints is higher in the shear mode than in the peel mode.

Three modes of failure are observed in this test: interfacial (IF), pullout (PF), and mixed (MF), which are depicted in Figure 4.40. The pullout mode yields the highest fracture load, and interfacial mode gives the lowest one. According to AWS D8.9M standard, the minimum value of the nugget size is equal to $4\sqrt{t}$ (t is the thickness of the base metal) [87], which according to [3] ensures a transition from IF to MF in spot welds of Al. The nugget size and the strength ratio of the nugget

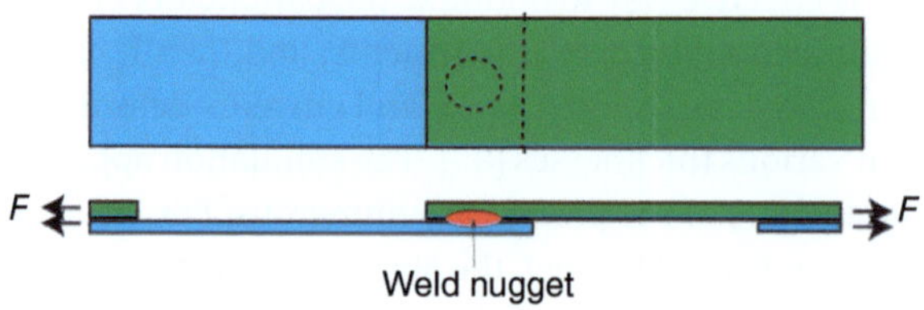

Figure 4.39 Tensile shear test for evaluation of the strength of the lap joints made by RSW.

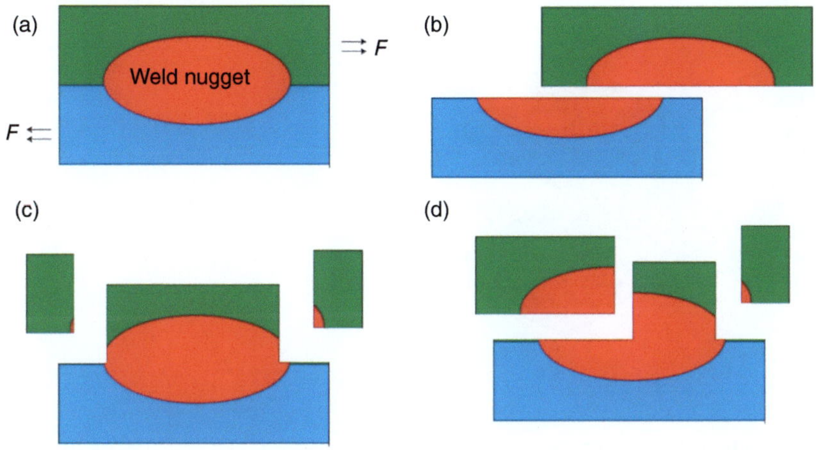

Figure 4.40 (a) Weld cross section. Three modes of failure of lap joints during tensile shear test. (b) IF, (c) PF, and (d) MF.

to its surrounding determine the mode of fracture. The minimum nugget size to ensure PF depends on the strength ratio. The HAZ of Al alloys is softened very high relative to the base material during RSW, making them to fail in the PF mode. But this PF mode of failure cannot guarantee a reasonable strength of the joint due to the very low strength of HAZ [68].

For calculation of the tensile stress of the lap joints, depending on whether the joint is failed in the PF or IF mode, the following equations are used:

$$\sigma_{PF} = \frac{Peak\ load}{2\pi rt} \tag{4.13}$$

$$\tau_{IF} = \frac{Peak\ load}{\pi r^2} \tag{4.14}$$

where σ_{PF} is the nominal peak joint stress for pullout mode, τ_{IF} is the nominal peak shear stress of the interfacial mode, r is the nugget radius, and t is the thickness of Al. The joint efficiency is defined as:

$$E_w = \frac{\sigma_{PF}\ or\ \tau_{IF}}{\sigma_{UTS}} \tag{4.15}$$

where σ_{UTS} is the ultimate tensile stress of Al.

Several empirical formulas have been developed to predict the fracture load (p) of the spot welds. Following equation is an example [3]:

$$p = C \cdot t \cdot d \cdot UTS \tag{4.16}$$

where C is a constant (equal to 3.1), t is the thickness of the base metal, d is the diameter of the button, and UTS is the ultimate tensile strength of the base metal. When the failure occurs in the IF mode, the strength of the joints mainly depends on the diameter of the nugget and is obtained from [88]:

$$p = C_{IF} \cdot UTS \cdot d^2 \tag{4.17}$$

where C_{IF} is a constant equal to 0.6.

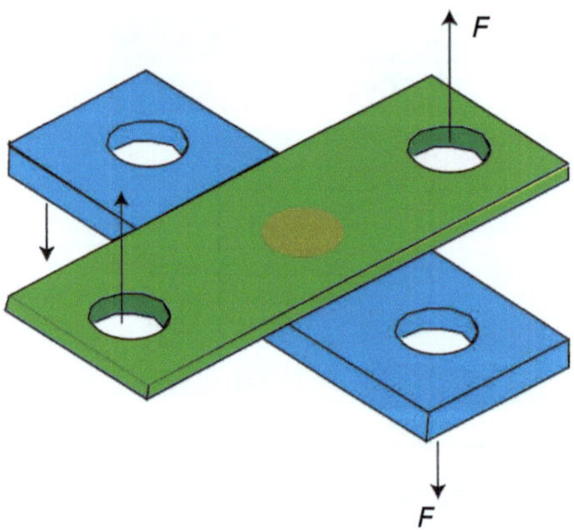

Figure 4.41 Schematic of cross-tension test.

When heat-treatable Al alloys are welded, an IF mode is observed due to a low hardness in the FZ caused by dissolution of precipitates. By a proper PWHT to retrieve the hardness, the failure mode changes into PF. In this state, the notch at the interface of the workpieces opens near the FZ and then turns into HAZ and PMZ due to the low hardness and strain concentration in these regions [24].

The other method named cross-tension test is used to evaluate the joint when loaded in normal direction to the interface (Figure 4.41). In this case, when the fracture is pullout, the peak load is obtained from [50]:

$$p = \pi t d \sigma_y \tag{4.18}$$

where σ_y is the YS of the fracture area. The main difference in the peak load formula for shear tensile and cross-tension test is the contribution of the YS in the peak load. This means by changing the microstructural characteristics that influence the YS, the cross-tension fracture load can be controlled [50].

Coach peeling is another testing method used to evaluate the mechanical properties of RSW joints. The coach peel coupon is comprised of two spot welds: the shunt weld and the test weld (Figure 4.42a). The shunt weld helps to prevent any additional distortion of the test coupon during the peeling test (Figure 4.42b).

Single-sided wedge testing is a method that is used to assimilate the coach peel testing. This method provides an in-situ observation of the crack initiation and propagation in coach-peel loading condition. In this method, a wedge at a specific rate is inserted to the interface of the two sheets jointed by RSW and a load-displacement curve is obtained. In load-displacement curve, a steady part exists that corresponds to the stage where the wedge is in contact with two sheets and bends them. Then, a peak is observed that corresponds to the contact of the wedge tip with the joint where the crack initiates (Figure 4.43) [89].

Fracture behavior of the joints made by RSW can be realized by FE simulation. The key points in such simulations are, first, to consider the material properties in each region of base, HAZ, and FZ and, second, to take into account the transition

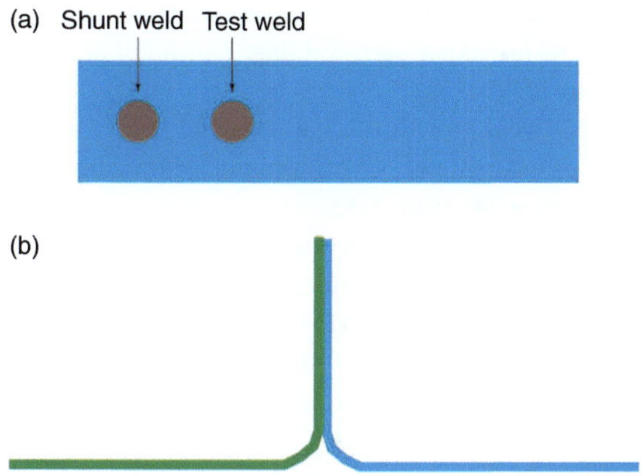

Figure 4.42 (a) Coach peel tensile test preparation. (b) Coach peel sample ready for testing.

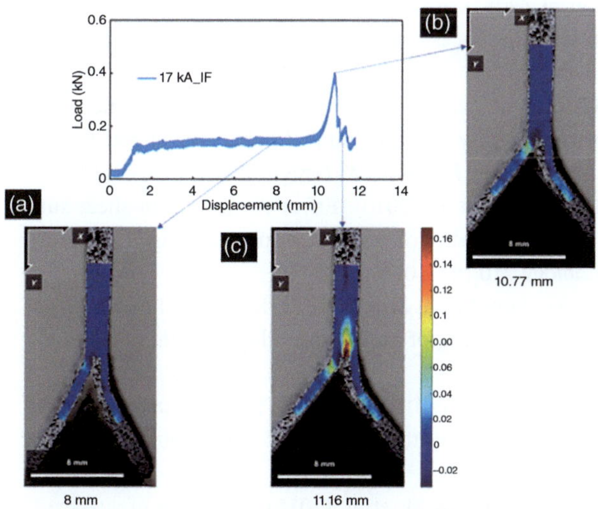

Figure 4.43 (a–c) Typical load-displacement curve obtained during a single-sided edge testing used for coach-peel testing of a RSW joint along with the situation of the wedge with respect to the joint. Source: Ref. [89]/© American Welding Journal.

layer between dissimilar materials. In dissimilar joints, the IMC layer properties are taken into account by approaches such as cohesive contact interface, as the thickness of the IMCs is very low (1–10 μm), and it is not realistic to explicitly model the IMC layer in macro-scale models [90].

References

1 Deepati, A.K., Alhazmi, W., and Benjeer, I. (2021). Mechanical characterization of AA5083 aluminum alloy welded using resistance spot welding for the lightweight automobile body fabrication. *Materials Today Proceedings* 45: 5139–5148.

2 Li, W., Cerjanec, D., and Grzadzinski, G.A. (2005). A comparative study of single-phase AC and multiphase DC resistance spot welding. *Journal of Manufacturing Science and Engineering* 127 (3): 583.
3 Li, Y., Yan, F., Luo, Z. et al. (2015). Weld growth mechanisms and failure behavior of three-sheet resistance spot welds made of 5052 aluminum alloy. *Journal of Materials Engineering and Performance* 24: 2546–2555.
4 Deng, L., Li, Y., Cai, W. et al. (2020). Simulating thermoelectric effect and its impact on asymmetric weld nugget growth in aluminum resistance spot welding. *Journal of Manufacturing Science and Engineering* 142 (9): 091001.
5 Epperlein, M., Schiebahn, A., Reisgen, U. et al. (2024). Joining aluminum die castings and wrought aluminum by resistance spot welding. *Journal of Materials: Design and Applications* 238 (12): 14644207241236388.
6 Bi, J., Song, J., Wei, Q. et al. (2016). Characteristics of shunting in resistance spot welding for dissimilar unequal-thickness aluminum alloys under large thickness ratio. *Materials & Design* 101: 226–235.
7 Ma, Y., Wu, P., Xuan, C. et al. (2013). Review on techniques for on-line monitoring of resistance spot welding process. *Advances in Materials Science and Engineering* 2013 (1): 630984.
8 Han, L., Thornton, M., Li, D. et al. (2011). Effect of governing metal thickness and stack orientation on weld quality and mechanical behaviour of resistance spot welding of AA5754 aluminium. *Materials & Design* 32 (4): 2107–2114.
9 Han, L., Thornton, M., Boomer, D. et al. (2010). Effect of aluminum sheet surface conditions on feasibility and quality of resistance spot welding. *Journal of Materials Processing Technology* 210 (8): 1076–1082.
10 Rashid, M., Fukumoto, S., Medley, J.B. et al. (2007). Influence of lubricants on electrode life in resistance spot welding of aluminum alloys. *Welding Research* 86 (3): 62.
11 Zhang, W., Cross, I., Fedman, P. et al. (2017). Electrode life of aluminum resistance spot welding in automotive applications: a survey. *Science and Technology of Welding and Joining* 22 (1): 22–40.
12 Al Naimi, I.K., Al Saadi, M.H., Daws, K.M. et al. (2015). Influence of surface pretreatment in resistance spot welding of aluminum AA1050. *Production & Manufacturing Research: An Open Access Journal* 3 (1): 185–200.
13 Zhang, Y., Shan, H., Li, Y. et al. (2017). Effects of the oxide film on the spot joining of aluminum alloy sheets: a comparative study between resistance spot welding and resistance spot clinching. *The International Journal of Advanced Manufacturing Technology* 92 (9): 4231–4240.
14 Dobosy, Á., Gáspár, M., and Török, I. (2018). Resistance spot welding of 7075 aluminium alloy. *Vehicle and Automotive Engineering 2: Proceedings of the Second VAE2018*, Miskolc, Hungary. Springer.
15 Hao, M., Osman, K.A., Boomer, D.R. et al. (1996). Developments in characterization of resistance spot welding of aluminum. *Welding Journal* 75 (1): 1–4.
16 Luo, Z., Ao, A., Chao, Y.J. et al. (2015). Application of pre-heating to improve the consistency and quality in AA5052 resistance spot welding. *Journal of Materials Engineering and Performance* 24: 3881–3891.

17 Lum, I., Biro, E., Yushan, Z. et al. (2004). Electrode pitting in resistance spot welding of aluminum alloy 5182. *Metallurgical and Materials Transactions A* 35: 217–226.

18 Zhou, Y., Fukumoto, S., Peng, J. et al. (2004). Experimental simulation of surface pitting of degraded electrodes in resistance spot welding of aluminium alloys. *Materials Science and Technology* 20 (10): 1226–1232.

19 Bamberg, P., Seewald, R., Schiebahn, A. et al. (2022). Improvement of the resistance spot welding of Al–Mg–Si alloys by using cladding technology: an optical and mechanical characterization study. *Journal of Advanced Joining Processes* 5 (1): 100090.

20 Rashid, M., Medley, J., and Zhou, Y. (2011). Nugget formation and growth during resistance spot welding of aluminium alloy 5182. *Canadian Metallurgical Quarterly* 50 (1): 61–71.

21 Kang, J., Chen, Y., Sigler, D. et al. (2015). Fatigue behavior of dissimilar aluminum alloy spot welds. *Procedia Engineering* 114: 149–156.

22 Wang, Y., Tao, W., and Yang, S. (2019). A method for improving joint strength of resistance spot welds of AA 5182-O aluminum alloy. *Journal of Manufacturing Processes* 45: 661–669.

23 Li, Y., Bi, J., Zhang, Y. et al. (2017). Shunting characteristics in triangular arranged resistance spot welding of dissimilar unequal-thickness aluminum alloys. *The International Journal of Advanced Manufacturing Technology* 91: 2447–2454.

24 Zhang, Y., Li, Y., Zhu, Z. et al. (2019). Bake-strengthening of resistance spot welded aluminum alloy 6061. *Welding Journal* 98: 337S–350S.

25 Kim, G.-C., Hwang, I., Kang, M. et al. (2019). Effect of welding time on resistance spot weldability of aluminum 5052 alloy. *Metals and Materials International* 25: 207–218.

26 Qi, L., Zhang, Q., Ma, Y. et al. (2021). A comparative study on mechanical performance of traditional and magnetically assisted resistance spot welds of A7N01 aluminum alloy. *Journal of Manufacturing Processes* 66: 133–144.

27 Deng, L., Li, Y., Carlson, B.E. et al. (2018). Effects of electrode surface topography on aluminum resistance spot welding. *Welding Journal* 97 (4): 120–132.

28 Rahimi, S. and Movahedi, M. (2020). Resistance spot welding of aluminum to aluminum clad steel sheet: experimental and theoretical analysis. *Journal of Manufacturing Processes* 58: 429–435.

29 Turnage, S., Darling, K.A., Rajagopalan, M. et al. (2018). Influence of variable processing conditions on the quasi-static and dynamic behaviors of resistance spot welded aluminum 6061-T6 sheets. *Materials Science and Engineering: A* 724: 509–517.

30 Wu, S., Ghaffari, B., Hetrick, E. et al. (2017). Thermo-mechanically affected zone in AA6111 resistance spot welds. *Journal of Materials Processing Technology* 249: 463–470.

31 Du, H., Bi, J., Zhang, Y. et al. (2019). The role of the partial melting zone in the nugget growth process of unequal-thickness dissimilar aluminum alloy

2219/5A06 resistance spot welding. *Journal of Manufacturing Processes* 45: 304–311.

32 Shi, L., Kang, J., Sigler, D.R. et al. (2019). Microstructure and fatigue behavior of novel Multi-Ring Domed resistance spot welds for thin X626-T4 aluminum sheets. *International Journal of Fatigue* 119: 185–194.

33 Li, H., Yan, W., Li, Z. et al. (2022). Numerical and experimental study of the hot cracking phenomena in 6061/7075 dissimilar aluminum alloy resistance spot welding. *Journal of Manufacturing Processes* 77: 794–808.

34 Senkara, J. and Zhang, H. (2000). Cracking in spot welding aluminum alloy AA5754. *Welding Journal* 79 (7): 194s–201s.

35 Zhang, H., Senkara, J., and Wu, X. (2002). Suppressing cracking in resistance welding AA5754 by mechanical means. *Journal of Manufacturing Science and Engineering* 124 (1): 79–85.

36 Wei, S., Zhang, R., Liu, X. et al. (2024). Comparative study on the effect of external magnetic field on aluminum alloy 6061 and 7075 resistance spot-welding joints. *Metals* 14 (10): 1196.

37 Hu, S., Haselhuhn, A.S., Ma, Y. et al. (2021). Sensitivity of dissimilar aluminum to steel resistance spot welds to weld gun deflection. *Journal of Manufacturing Processes* 68: 534–545.

38 Brozek, M. (2015). Resistance spot welding of steel sheets of different thickness. *Engineering for Rural Development*, Jelgava (20–22 May 2015), pp. 72–77. Latvia University of Agriculture.

39 Wang, B., Qiu, F., Chen, L. et al. (2022). Microstructure and shearing strength of stainless steel/low carbon steel joints produced by resistance spot welding. *Journal of Materials Research and Technology* 20: 2668–2679.

40 Lee, T. (2020). Resistance spot weldability of heat-treatable and non-heat-treatable dissimilar aluminium alloys. *Science and Technology of Welding and Joining* 25 (7): 543–548.

41 Qin, Q., Zhao, H., Zhang, Y. et al. (2019). Microstructures and mechanical properties of Al–Mg_2Si–Si alloys resistance spot welded with Al–Si interlayers. *Journal of Materials Research and Technology* 8 (5): 4318–4332.

42 Zedan, M.J. and Doos, Q.M. (2018). New method of resistance spot welding for dissimilar 1008 low carbon steel-5052 aluminum alloy. *Procedia Structural Integrity* 9: 37–46.

43 Chen, N., Wang, M., Wang, H.-P. et al. (2018). Microstructural and mechanical evolution of Al/steel interface with Fe_2Al_5 growth in resistance spot welding of aluminum to steel. *Journal of Manufacturing Processes* 34: 424–434.

44 Shi, L., Kang, J., Qian, C. et al. (2022). Role of Fe_2Al_5 in fracture of novel dissimilar aluminum-steel resistance spot welds using multi-ring domed electrodes. *Materials Science and Engineering: A* 831: 142233.

45 Hu, S., Haselhuhn, A.S., Ma, Y. et al. (2020). Comparison of the resistance spot weldability of AA5754 and AA6022 aluminum to steels. *Welding Journal* 99 (8): 224S–238S.

46 Che, Y., Wang, L., Sun, D. et al. (2018). Microstructures and mechanical properties of resistance spot-welded steel/aluminum alloy joints with process tapes. *Materials & Design* 27: 5532–5544.

47 Pan, B., Sun, H., Shang, S.-L. et al. (2022). Understanding formation mechanisms of intermetallic compounds in dissimilar Al/steel joint processed by resistance spot welding. *Journal of Manufacturing Processes* 83: 212–222.

48 Son, B., Woo, S.Y., Jb, L. et al. (1997). A study on resistance spot welding of dissimilar sheet metals (aluminium alloy-steel sheets). *Journal of KWS* 15 (4): 42–62.

49 Das, T. and Paul, J. (2020). Resistance spot welding of similar and dissimilar metals: the effect of graphene interlayer. *Journal of Metals* 72: 2863–2874.

50 Li, M., Tao, W., Zhang, J. et al. (2022). Hybrid resistance-laser spot welding of aluminum to steel dissimilar materials: microstructure and mechanical properties. *Materials & Design* 221: 111022.

51 Chen, C., Kong, L., Wang, M. et al. (2019). The robustness of Al-steel resistance spot welding process. *Journal of Manufacturing Processes* 43: 300–310.

52 Zhang, Y., Li, Y., Luo, Z. et al. (2016). Feasibility study of dissimilar joining of aluminum alloy 5052 to pure copper via thermo-compensated resistance spot welding. *Materials & Design* 106: 235–246.

53 Zare, M. and Pouranvari, M. (2021). Metallurgical joining of aluminium and copper using resistance spot welding: microstructure and mechanical properties. *Science and Technology of Welding & Joining* 26 (6): 461–469.

54 Alikhani, A., Beygi, R., Mehrizi, M.Z. et al. (2021). Effect of Mg and Si on intermetallic formation and fracture behavior of pure aluminum-galvanized carbon-steel joints made by weld-brazing. *Journal of Central South University* 28 (11): 3626–3638.

55 Yagati, K.P., Bathe, R.N., Rajulapati, K.V. et al. (2014). Fluxless arc weld-brazing of aluminium alloy to steel. *Journal of Materials Processing Technology* 214 (12): 2949–2959.

56 Tan, C., Zang, C., Xia, H. et al. (2018). Influence of Al additions in Zn–based filler metals on laser welding–brazing of Al/steel. *Journal of Manufacturing Processes* 34: 251–263.

57 Zhang, W., San, D., Han, L. et al. (2011). Characterization of intermetallic compounds in dissimilar material resistance spot welded joint of high strength steel and aluminum alloy. *ISIJ International* 51 (11): 1870–1877.

58 Lu, Y., Mayton, E., Song, H. et al. (2019). Dissimilar metal joining of aluminum to steel by ultrasonic plus resistance spot welding-microstructure and mechanical properties. *Materials & Design* 165: 107585.

59 Enrique, P.D., Li, C., DiGiovanni, D. et al. (2020). Electrospark deposition interlayers for dissimilar resistance welding of steel to aluminum. *Manufacturing Letters* 24: 123–126.

60 Winnicki, M., Małachowska, A., Korzeniowski, M. et al. (2018). Aluminium to steel resistance spot welding with cold sprayed interlayer. *Surface Engineering* 34 (3): 235–242.

61 Zhang, G., Zhao, H., Xu, X. et al. (2019). Metallic bump assisted resistance spot welding (MBaRSW) of AA6061-T6 and bare DP590: part I-printing of metallic bump. *Journal of Manufacturing Processes* 44: 427–434.

62 Zhang, G., Zhao, H., Xu, X. et al. (2019). Metallic bump assisted resistance spot welding (MBaRSW) of AA6061-T6 and Bare DP590: part II-joining mechanism and joint property. *Journal of Manufacturing Processes* 44: 19–27.

63 Das, T., Das, R., and Paul, J. (2020). Resistance spot welding of dissimilar AISI-1008 steel/Al-1100 alloy lap joints with a graphene interlayer. *Journal of Manufacturing Processes* 53: 260–274.

64 Lu, Y., Walker, L., Kimchi, M. et al. (2020). Microstructure and strength of ultrasonic plus resistance spot welded aluminum alloy to coated press hardened boron steel. *Metallurgical and Materials Transactions A* 51: 93–98.

65 Qiu, R., Li, J., Shi, H. et al. (2023). Characterization of resistance spot welded joints between aluminum alloy and mild steel with composite electrodes. *Journal of Materials Research and Technology* 24: 1190–1202.

66 Maddela, S. and Carlson, B.E. (2019). Corrosion characterization of resistance spot-welded aluminum and steel couple. *Journal of Manufacturing Science and Engineering* 141 (11): 111010.

67 Zhang, Y., Shan, H., Li, Y. et al. (2017). Joining aluminum alloy 5052 sheets via novel hybrid resistance spot clinching process. *Materials & Design* 118: 36–43.

68 Zhang, Y., Zhang, X., Guo, J. et al. (2019). Effects of local stiffness on the spot joints mechanical properties: comparative study between resistance spot welding and resistance spot clinching joints. *Journal of Manufacturing Processes* 39: 93–101.

69 Zhao, D., Ren, D., Song, G. et al. (2021). Nugget formation analysis of Al/Steel clinch-resistance hybrid spot welding. *Science and Technology of Welding and Joining* 26 (6): 439–447.

70 Shah, U. and Liu, X. (2020). Effect of ultrasonic energy on the spot weldability of aluminum alloy AA6061. *Materials & Design* 192: 108690.

71 Bamberg, P., Gintrowski, G., Liang, Z. et al. (2021). Development of a new approach to resistance spot weld AW-7075 aluminum alloys for structural applications: an experimental study–Part 1. *Journal of Materials Research and Technology* 15: 5569–5581.

72 Huang, M., Zhang, Q., Qi, L. et al. (2020). Effect of external magnetic field on resistance spot welding of aluminum alloy AA6061-T6. *Journal of Manufacturing Processes* 50: 456–466.

73 Eder, A., S. Jaber, and N. Jank, Using simulation for investigations of hot cracking phenomena in resistance spot welding of 6xxx aluminum alloys (AA6016 and AA6181). In *Hot Cracking Phenomena in Welds II*, T. Bollinghaus, H. Herold,·C.E. Cross, J.C. Lippold (Eds.). 2008, Springer. p. 311–325.

74 Shin, S., Park, D.-J., Yu, J. et al. (2019). Resistance spot welding of aluminum alloy and carbon steel with spooling process tapes. *Metals* 9 (4): 410.

75 Kim, Y.-G., Kim, D.-C., Joo, S.-M. et al. (2019). Evaluation of tensile shear strength for dissimilar spot welds of Al–Si–Mg aluminum alloy and galvanized steel by delta-spot welding process. *Journal of Mechanical Science and Technology* 33: 5399–5405.

76 Feng, Y., Luo, Z., Li, Y. et al. (2016). A novel method for resistance plug welding of 7075 aluminum alloy. *Materials and Manufacturing Processes* 31 (16): 2077–2083.

77 Shah, U. and Liu, X. (2019). Effects of ultrasonic vibration on resistance spot welding of transformation induced plasticity steel 780 to aluminum alloy AA6061. *Materials & Design* 182: 108053.

78 Florea, R., Hubbard, C.R., Solanki, K.N. et al. (2012). Quantifying residual stresses in resistance spot welding of 6061-T6 aluminum alloy sheets via neutron diffraction measurements. *Journal of Materials Processing Technology* 212 (11): 2358–2370.

79 Nielsen, C.V., P.A.F. Martins, W. Zhang et al. Numerical methods in simulation of resistance welding. *COUPLED VI: Proceedings of the VI International Conference on Computational Methods for Coupled Problems in Science and Engineering*, Venice, Italy (18–20 May 2015). CIMNE.

80 Beygi, R., E. Marques, and L.F. da Silva, Computational Concepts in Simulation of Welding Processes. 2022: Springer.

81 Karimi, M., Sedighi, M., and Afshari, D. (2015). Thermal contact conductance effect in modeling of resistance spot welding process of aluminum alloy 6061-T6. *The International Journal of Advanced Manufacturing Technology* 77: 885–895.

82 Sun, X. and Dong, P. (2000). Analysis of aluminum resistance spot welding processes using coupled finite element procedures. *Welding Journal* 79 (8): 215S.

83 Afshari, D., Sedighi, M., Karimi, M.R. et al. (2016). Prediction of residual stresses in resistance spot weld. *Aircraft Engineering and Aerospace Technology* 88 (4): 492–497.

84 Sedighi, M., Afshari, D., and Nazari, F. (2018). Investigation of the effect of sheet thickness on residual stresses in resistance spot welding of aluminum sheets. *ARCHIVE Proceedings of the Institution of Mechanical Engineers, Part C: Journal of Mechanical Engineering Science* 232 (4): 621–638.

85 Schulz, E., Wagner, M., Schubert, H. et al. (2021). Short-pulse resistance spot welding of aluminum alloy 6016-T4 – Part. *Welding Journal* 100: 41–51.

86 Zhou, K., Ren, B., and Yu, W. (2023). Optimized designing of generalized electrodes for aluminum/steel resistance spot welding process based on numerical calculation. *Journal of Manufacturing Processes* 99: 563–580.

87 American Welding Society (2022). *Test Methods for Evaluating the Resistance Spot Welding Behavior of Automotive Sheet Steel Materials*. American Welding Society.

88 Radakovic, D. and Tumuluru, M. (2008). Predicting resistance spot weld failure modes in shear tension tests of advanced high-strength automotive steels. *Welding Journal* 87 (4): 96.

89 Lu, Y., Mayton, E., Song, H. et al. (2019). Optimizing ultrasonic plus resistance spot welding for dissimilar metal joining. *Materials & Design* 98: 273S–282S.

90 Chen, J., Feng, Z., Wang, H.-P. et al. (2018). Multi-scale mechanical modeling of Al-steel resistance spot welds. *Materials Science and Engineering: A* 735: 145–153.

5 Solid-State Welding Process

5.1 Introduction

FSW is one of the most sustainable welding processes due to its low energy consumption and absence of filler metals or inert gases [1]. When applied to Al, FSW largely avoids the major defects typically associated with fusion welding – such as SC, liquation cracking, and softening – thanks to its lower heat input and absence of fusion. However, FSW may introduce technical defects like cavities and flash, which can be mitigated through the careful selection of tool geometry and process parameters [2]. A key advantage of FSW is that it does not require pre-treatment, such as edge preparation or the use of filler materials, thus reducing overall process costs. Furthermore, FSW is particularly effective for welding Al-matrix composites, which are challenging to weld with fusion methods due to severe porosity, particle agglomeration, and the formation of IMCs from reactions between reinforcement particles and the melt during fusion processes [3].

FSW is performed in various weld designs such as butt, lap, and Tee configurations. One disadvantage of FSW in lap joint configurations, compared to fusion welding, is reduced strength on the retreating side, outside the stir zone [4]. This weakness arises from hook formation, which creates an unbonded region, reducing the effective load-bearing thickness [5]. However, in butt joint configurations, FSW shows clear advantages over fusion welding, producing superior welds primarily due to the homogeneous, fine microstructure developed in the weld zone [6].

Since its invention by The Welding Institute (TWI) in 1991, numerous modifications and advancements have been made to address FSW's limitations. One significant limitation is the need for backing support and fixturing tools to secure the plates, which has been mitigated by the development of bobbin tool FSW, eliminating the requirement for a support plate [7]. Despite these advancements, the fundamental principles of joint formation remain consistent across FSW variants, relying on the formation of a metallurgical bond. However, when joining dissimilar materials to Al, precise control is essential to manage the formation of brittle IMCs [8]. The performance of these joints is highly dependent on the thickness and morphology of the IMCs, underscoring the importance of investigating process parameters to control IMC formation [9]. FSW has also been adapted to create mechanical bonds through macro-interlocking of materials via plastic deformation. In these

Welding Metallurgy of Aluminium Alloys: Design, Processes, and Simulations, First Edition.
Reza Beygi, Eduardo André de Sousa Marques, Ricardo João Camilo Carbas, and Lucas Filipe Martins da Silva.

joints, metallurgical bonding is not the primary mechanism, and they are thus not classified as welding [10]. Additionally, FSW can be combined with other joining techniques, such as adhesive bonding, though the distinct bonding mechanisms of these methods require careful design to minimize potential interference [11].

This chapter focuses on the conventional FSW process, which relies on metallurgical bonding. It explores microstructural evolution during FSW and its impact on joint mechanical performance. Various joint designs, their feasibility, and performance are discussed, emphasizing how microstructural changes influence mechanical properties. Mechanical joints formed solely through plastic deformation are not covered, as the book centers on welding metallurgy, where metallurgical bonding is essential.

5.2 FSW of Al Alloys

After FSW of Al alloys, three distinct regions are observed: the NZ, TMAZ, and HAZ [12]. The defining feature of these zones, visible under optical microscopy, is the variation in grain structure and size. The NZ itself can be further divided into distinct regions, though no clear boundary exists between them. This variation in grain size from the top to the bottom of the NZ is attributed to two factors [13]: the higher heat input at the top, due to the shoulder's influence, and the heat sink effect of the backing plate at the bottom, which leads to lower temperatures. As a result, the grain structure at the top of the NZ is coarser than at the bottom. The primary mechanism of grain refinement within the NZ is dynamic recrystallization, and the grain size after dynamic recrystallization ($\overline{D}$) can be estimated using the following equation [14]:

$$\overline{D}^2 - \overline{D_0}^2 = At \exp\left(\frac{-Q}{RT}\right) \tag{5.1}$$

where $\overline{D}_0$ is the initial grain size, A is a constant, Q is the activation energy, R is the gas constant, and T is the temperature.

To study the microstructural changes in Al alloys during FSW, it is essential to distinguish between HTA and non-heat-treatable (NHTA) Al alloys. The primary strengthening mechanisms in NHTAs are solid-solution hardening and strain hardening, whereas precipitation strengthening is the dominant mechanism in HTAs [15]. This distinction leads to notable differences in the FSW process for these two groups of alloys.

In both HTAs and NHTAs, softening occurs in the NZ, TMAZ, and HAZ, though the underlying mechanisms differ. In NHTAs, softening results from dynamic recrystallization, which reduces the density of dislocations that had previously been introduced by strain hardening – their primary strengthening mechanism. In HTAs, in addition to recrystallization and grain refinement, precipitate changes, such as dissolution or coarsening, occur depending on the alloy and the thermal cycle during FSW.

While grain changes in these zones are visible under optical microscopy, precipitate changes typically require more advanced techniques, such as SEM or TEM. These microstructural changes are also evident in the hardness profile of the weld.

In HTAs, strength can be restored through PWHT. However, in NHTAs, strength recovery through PWHT is not possible for two main reasons:

1) Al alloys, unlike carbon steels, do not undergo phase transformations in the solid state, which means they cannot be heat-treated to recover strength, such as through normalizing.
2) The absence of precipitates in NHTAs means they cannot retain or recover strength through solution treatment and aging processes.

The following section discusses the specific changes that occur after PWHT in HTAs. The contributions of grain size and precipitates to the increase of the strength ($\Delta\sigma$) of the Al alloys are obtained through the following equations, respectively [16]:

$$\Delta\sigma_{grain\ size} = kd^{-0.5} \tag{5.2}$$

$$\Delta\sigma_{percipitate} = \frac{0.13Gb}{\lambda} \ln \frac{r}{b} \tag{5.3}$$

where k is a constant, d is the grain size, b is the Burgers vector, r is the radius of the precipitates, and λ is the space between the precipitates. In Al-matrix composites, the reinforcing particles also influence the microstructure of the NZ during FSW. Fine reinforcing particles promote the nucleation of recrystallized grains by inducing lattice misorientation and accumulating local strain around the particles, resulting in finer grain sizes in the NZ [3].

The final strength of the NZ is determined by the interplay of three main phenomena during FSW: grain refinement, dynamic recrystallization, and precipitate dissolution. In heat-treatable Al alloys, the NZ often experiences a significant loss of hardness and strength after FSW, primarily due to the dissolution of precipitates. To recover this strength, a PWHT consisting of solution treatment followed by aging is required. While PWHT improves joint strength, it often leads to a reduction in ductility in the NZ due to grain coarsening (Figure 5.1a, b) [17]. The increase in grain size during PWHT results in decreased ductility, but the hardness of the NZ increases despite the grain growth. This suggests that precipitate hardening is the dominant factor determining the joint strength after PWHT of HTAs.

A PWHT consisting only of an aging treatment can partially recover the hardness in the NZ. The extent of this recovery depends on the degree of solutioning that

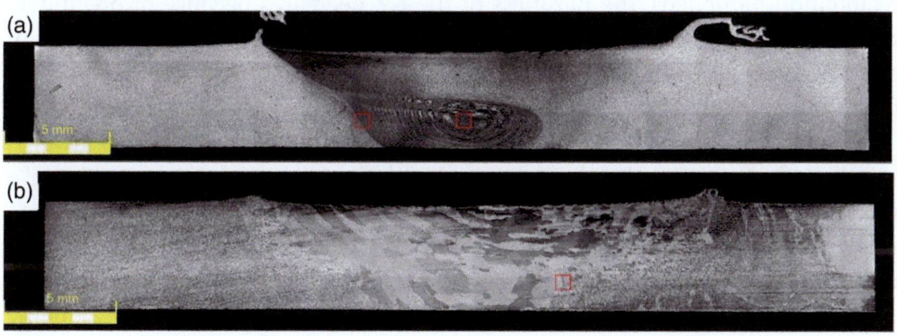

Figure 5.1 Macro-image of an FSW weld made on an AA2519 (a) after welding and (b) after PWHT (solutioning + aging). Source: Ref. [18]/MDPI/CC BY 3.0.

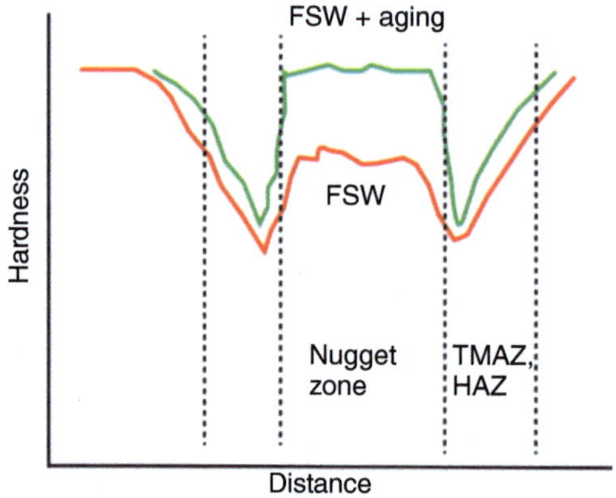

Figure 5.2 Schematic of hardness profile of a heat-treatable Al after FSW (as-welded) and after PWHT (only aging).

occurred during FSW. The results of the effect of PWHT (aging for 12 hours) on the increase in tensile strength and hardness in the NZ of an AA6082 alloy show that the increase in properties is more pronounced in samples welded at lower welding speeds compared to those welded at higher speeds. This difference is due to the greater degree of solutioning in the NZ of the samples welded at lower speeds, which results from the higher heat input during welding [19]. As a result, more solute elements are retained in the NZ, providing more activation energy for precipitation during the PWHT, leading to greater strength and hardness recovery.

One advantage of using only an aging treatment for PWHT is that the grains in the NZ do not grow, due to the lower temperatures involved. However, the hardness in the HAZ and TMAZ cannot be restored because these regions experience precipitate coarsening with minimal solutioning during FSW. As a result, the driving force for precipitation in these areas is very low, and they do not respond effectively to aging. Figure 5.2 illustrates how the hardness in the NZ is partially recovered after PWHT (aging only), while the hardness in the HAZ and TMAZ remains unchanged [19]. In contrast, a full hardness recovery is observed after a PWHT that includes both solutioning and aging, resulting in a uniform hardness profile.

Another factor influencing the hardness profile after PWHT is the initial temper condition of the Al alloy. When a heat-treatable Al alloy is welded in the O-temper condition (fully solutionized), FSW leads to an increase in hardness in the NZ due to grain refinement. In contrast, when the alloy is welded in the T6 condition (precipitation-hardened), the hardness of the NZ decreases because of precipitate dissolution. In the O-temper condition, grain size is the dominant factor, while in the T6 condition, precipitates play a more significant role. Therefore, the hardness profile of an FSWed Al alloy in the O-temper condition shows a rise in the NZ, as strengthening follows the Hall–Petch relationship, where grain size governs strength [20]. A PWHT that includes both solutionizing and aging results in a uniform hardness profile in both O-temper and T6-temper samples [21].

Though solution treatment and aging are favored over aging as they produce a uniform hardness profile, they are associated with a phenomenon that disturbs the uniformity of the grain size in the NZ [22]. Despite the advantage of producing a uniform hardness profile, solution treatment followed by aging can lead to

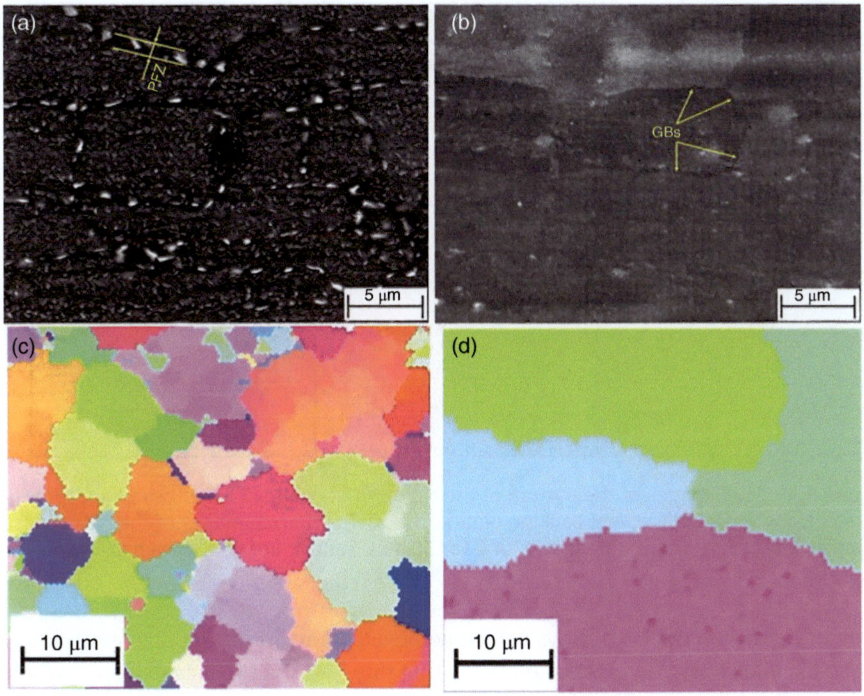

Figure 5.3 Precipitation morphologies in (a) air-cooled and (b) water-cooled NZ after solutioning. EBSD image of the NZ after PWHT of (c) air-cooled and (d) water-cooled NZ. Source: Ref. [24]/Springer Nature.

abnormal grain growth in the NZ during PWHT [23]. The primary consequence of abnormal grain growth is a reduction in ductility. To understand the impact of grain boundary pinning on abnormal grain growth in the NZ during PWHT, Figure 5.3a and b shows two precipitation morphologies, corresponding to air-cooled and water-cooled NZs after solution heat treatment. During air cooling, solutes have sufficient time to diffuse to grain boundaries and nucleate precipitates. These precipitates grow, depleting the surrounding grain boundary regions and forming a precipitate-free zone (PFZ) during aging. The pinning effect of these precipitates at grain boundaries inhibits abnormal grain growth (Figure 5.3c). In contrast, in the water-cooled NZ, the absence of precipitates at grain boundaries results in abnormal grain growth (Figure 5.3d) [24].

Abnormal grain growth results from secondary recrystallization, driven by a fine-grained structure. This phenomenon occurs when the thermodynamic driving force for grain growth exceeds the pinning force at grain boundaries. In the NZ, where precipitates have dissolved during FSW, abnormal grain growth becomes more likely due to the lack of precipitate-induced pinning [25].

The grain size after FSW significantly affects grain growth during PWHT. At the bottom of the NZ, where the backing plate acts as a heat sink, the cooling rate is the highest, resulting in a finer grain structure. This region is therefore more susceptible to grain growth during PWHT [26]. Controlling abnormal grain

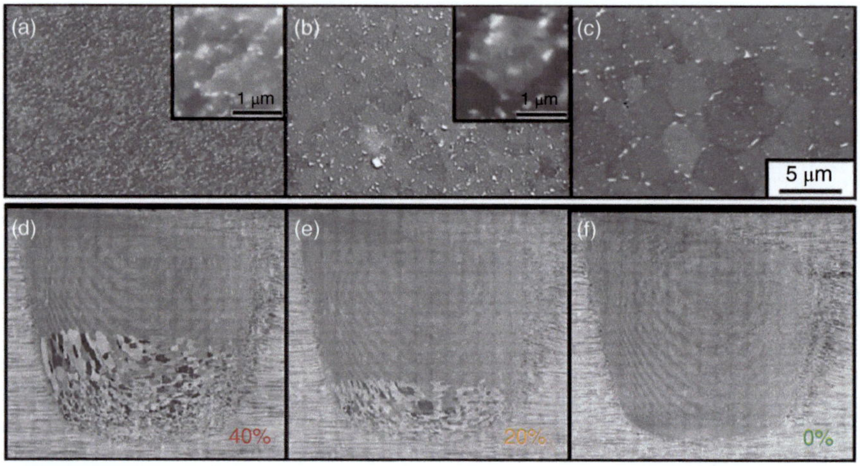

Figure 5.4 SEM images of the bottom of the NZ of AA7xxx alloy FSWed using (a) copper, (b) steel, and (c) stainless steel backing plates. (d, e, f) The macro-images of the NZ of these samples after solution treatment at 470 °C for 30 min. Source: Ref. [26]/with permission of Elsevier.

growth in FSWed AA7xxx alloys can be achieved using a backing plate with lower thermal conductivity to moderate the cooling rate [26]. Figure 5.4a–c displays SEM images of the bottom of the NZs of AA7xxx alloy FSWed using three different backing plates: copper, steel, and stainless steel, respectively. A larger grain size is observed when using the stainless steel backing plate (Figure 5.4c) due to its lower heat conductivity, which encourages grain growth during FSW. When this sample undergoes PWHT, no abnormal grain growth is observed, as the larger grain size provides less driving force for further grain growth. This relationship is explained by the following equation [27]:

$$\frac{d\overline{D}}{dt} = \alpha M \frac{2\gamma}{\overline{D}} \tag{5.4}$$

where $\overline{D}$ is the average grain size, t is time, α is a constant, γ is the surface energy of the grain boundary, and M is the mobility of the grain boundary.

For recovering the strength of NHTAs after FSW, alternative approaches to PWHT must be considered. The main NHTA groups include AA1xxx, AA3xxx, and AA5xxx series, typically designated by the temper code "H," followed by two or more digits that indicate the level of strain hardening. When these alloys are in the annealed condition, the temper designation "O" is used [28]. In this state, they exhibit the lowest strength, which is primarily achieved through solid-solution hardening. Figure 5.5 presents the hardness profile of a weld in an AA5083 alloy. A noticeable reduction in hardness can be observed in both the NZ and HAZ, along with an asymmetric hardness profile. This asymmetry is attributed to the differing thermal and strain histories between the retreating and advancing sides during FSW [29].

Since PWHT is not applicable for NHTAs, the FSW process must be carefully controlled or modified to minimize the heat effect on the weld. One approach is to reduce the heat input by using lower tool rotation speeds during FSW [30]. However,

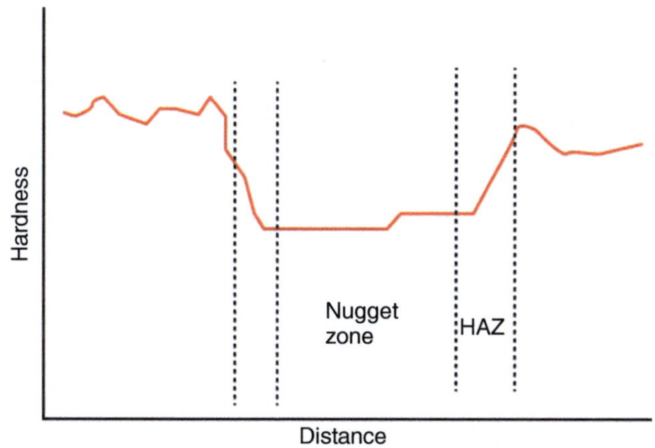

Figure 5.5 Schematic of hardness profile of the weld of a non-heat-treatable Al.

there is a lower boundary for rotation speed, as insufficient material flow at low heat input can lead to defects such as tunneling. Figure 5.6 shows schematic of hardness profiles of an AA1xxx alloy welded at various tool rotation speeds. The profile changes from a U-shape at high rotation speeds to a reverse U-shape at low rotation speeds. At high rotation speeds, full recrystallization occurs, leading to the elimination of dislocations and the formation of sub-grains. Conversely, at low rotation speeds, recrystallization is incomplete, and the dislocation density remains high. One significant observation in Figure 5.6 is the increase in hardness of the NZ compared to the base material in the sample welded at low rotation speeds, despite the base material being in a cold-rolled condition. This increase is due to the plastic deformation induced by the tool pin, which, in combination with the low temperature, raises the dislocation density relative to the base material.

Another approach to control the microstructure of NHTAs during FSW is to use a cooling medium, such as underwater FSW. For HTAs, using water as a cooling medium helps to reduce precipitate coarsening in the TMAZ and decrease the width

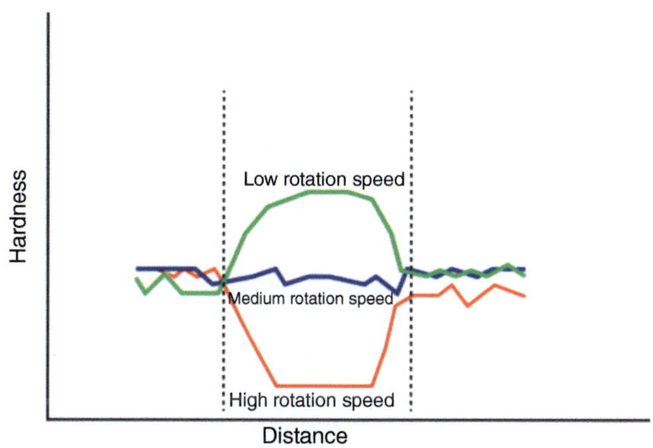

Figure 5.6 Schematic of hardness profile of a weld of a non-heat-treatable Al FSWed at various rotation speeds of the tool.

of the TMAZ, leading to a more uniform hardness profile across the weld. Additionally, it reduces the grain size in the NZ, which enhances the mechanical performance of the weld [31]. In the case of NHTAs, the high cooling rate and lower peak temperatures achieved during underwater FSW help to prevent grain growth during recrystallization in the NZ [32]. When the alloy is in a work-hardened (H-temper) condition, a decrease in hardness is typically observed in the NZ after FSW due to the elimination of dislocations. However, when FSW is performed underwater, the hardness of the NZ can increase relative to the base metal due to the formation of a very fine grain structure (Figure 5.7). A different behavior is observed when the alloy is annealed before FSW (O-temper). In this case, the hardness of the NZ increases relative to the base material even when there is not a cooling medium, because dislocations are absent in the annealed base material. The Hall–Petch equation becomes the primary factor in determining strength, as the grain size in the NZ after FSW is significantly smaller than that of the base material (Figure 5.7) [33].

Another method to enhance the strength in the NZ of NHTAs involves using reinforcements within the NZ, similar to the process occurring during FSW of Al-matrix composites, which will be discussed in a later section. The main challenges with this approach are the nonuniform dispersion of particles in the NZ and the inability to incorporate these reinforcements into the HAZ and TMAZ [34].

The hardness profiles discussed above apply to welds with thicknesses below a specific value, typically around 20 mm. In thicker Al alloys, referred to as "thick" Al alloys, the microstructural evolution is not uniform from the top to the bottom of the weld. The top of the weld is influenced by the shoulder, receiving more heat, while the bottom is more affected by the plastic deformation caused by the pin. As a result, the shoulder-affected zone at the top of the NZ in thick plates experiences grain growth and dislocation elimination, leading to a decrease in strength. Deeper into the NZ, the material undergoes significant plastic deformation from the pin while receiving less heat, resulting in a high dislocation density within the grains and along sub-grain boundaries [35]. Figure 5.8 presents a schematic of the hardness profile in the weld region of a thick NHTA (5A06) after FSW. In the shoulder-affected zone, the hardness profile exhibits a U-shape due to strength loss from the shoulder's heat.

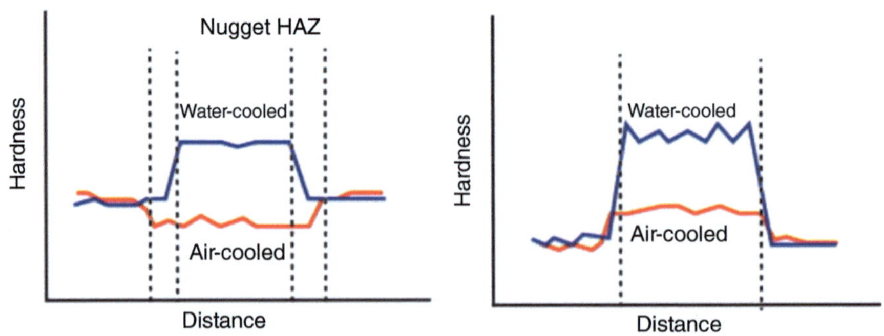

Figure 5.7 Schematic of hardness profiles of the weld of a non-heat-treatable Al cooled in air and water when the initial temper of Al is hardened (H-temper), and annealed (O-temper).

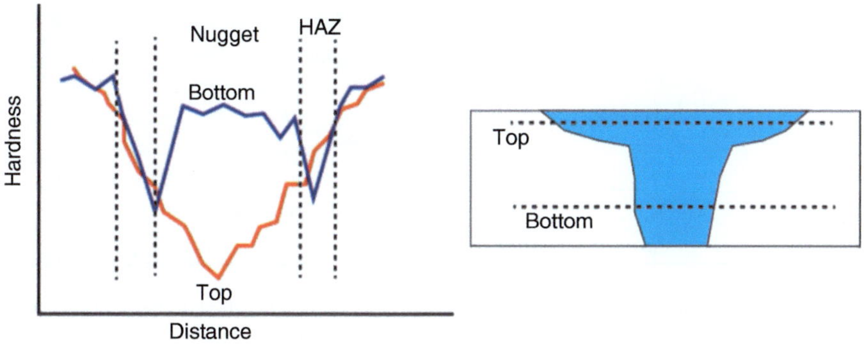

Figure 5.8 Schematic of hardness distribution in the weld of a thick non-heat-treatable Al, measured at top and bottom.

TEM image of this region reveals the formation of sub-boundaries with a low dislocation density. Further down into the NZ, the hardness profiles shift to a W-shape, attributed to the strain-hardening effect of the pin and insufficient heat from the shoulder to induce full recrystallization and grain growth. TEM images of this region show a high density of dislocations. The reduction in hardness in the HAZ is caused by softening and the absence of plastic deformation from the pin.

Various heat treatments, such as solution treatment, natural aging, and artificial aging after FSW, produce different hardness profiles, which ultimately determine the weakest location during tensile testing [36]. In a defect-free joint, failure typically occurs in the region with the lowest hardness. Beyond the microstructure, defects also play a crucial role in influencing mechanical properties. For instance, the pin height is often shorter than the sheet thickness, which can lead to the formation of a root defect or lack of penetration if there is insufficient heat and plastic deformation in butt welding. This defect degrades the joint's performance in mechanical tests but can be eliminated by optimizing the process parameters [37]. Kissing bonds are another significant defect that occurs when inadequate heat input and plastic flow fail to disrupt the oxide layers present on the initial surfaces of the sheets. Figure 5.9 shows an example of a lack of penetration defects during FSW of two dissimilar alloys of Al. These defects not only lower the joint's strength but also result in brittle fracture behavior during tensile testing (Figure 5.9a) [38]. In a defect-free joint, failure occurs in the softest region, exhibiting high ductility, as evidenced by the obvious necking at the fracture site. The brittleness of the joints with these defects is apparent, as there is minimal thickness reduction at the fracture locations. Any potential improvement in joint strength by modifying the microstructure through PWHT becomes ineffective in the presence of these defects [25]. SEM fractography further confirms the brittle characteristics when a failure occurs via kissing bonds [39].

During welding of dissimilar Al alloys, a distinct boundary is typically observed between the two alloys when using a tapered or cylindrical pin (as is observed in Figure 5.9). However, employing a pyramidal-shaped pin results in an NZ consisting of alternating layers of the two Al alloys, often accompanied by voids. These voids

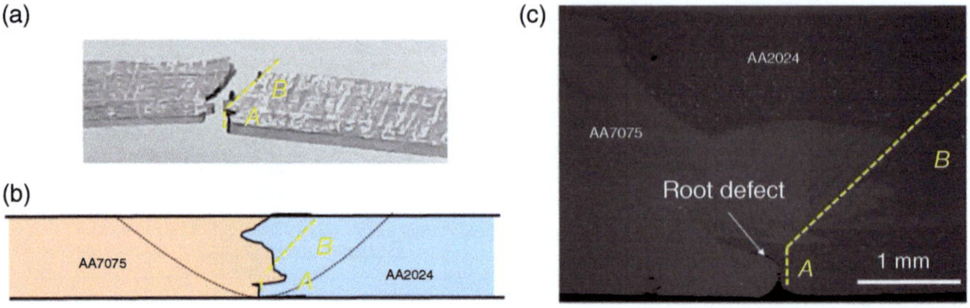

Figure 5.9 (a) Fracture in an FSW weld with lack of penetration. (b) Schematic of the weld cross section made between two Al alloys. (c) SEM image showing the lack of penetration. Source: Ref. [40]/MDPI/CC BY 4.0.

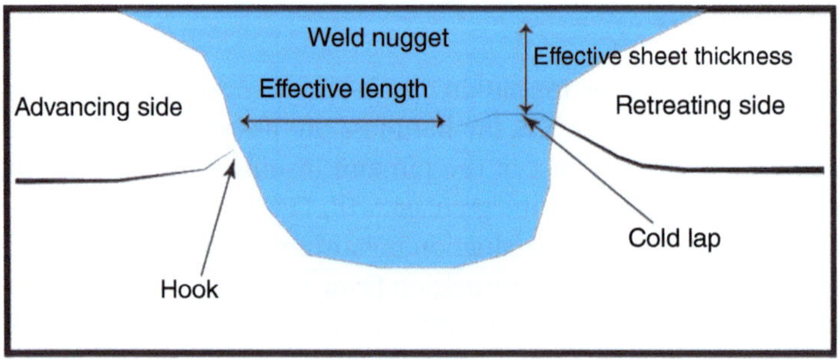

Figure 5.10 Schematic of cold lap and hook formed during FSW in lap configuration.

arise due to the differing flow characteristics of the alloys, which lead to inconsistencies in the material flow between the alternating layers [40].

Cold lap is a defect that occurs during FSW in lap joint configurations. It is caused by the lifting of the interface due to upward material flow, which reduces the load-bearing capacity of the joint, as the lifted interface lacks a proper metallurgical bond [41]. A deep plunging of the tool exacerbates this defect. Figure 5.10 illustrates this defect, showing how the effective sheet thickness and effective lap width are compromised by the presence of a cold lap. Another defect, known as a hook, is also shown in Figure 5.10. Similar to the cold lap, the hook is lifted upward, but with the key difference that it forms in TMAZ and disappears within the weld nugget [42].

Other than mechanical properties, the corrosion characteristics are also affected by microstructure evolution during FSW through grain size and precipitation coarsening [43]. This property needs to be investigated when needed.

5.3 FSW of Al-Matrix Composites

Reinforcing particles in Al-matrix composites offer an advantage over precipitates, as they do not grow or dissolve at high temperatures, whether during service or processing such as FSW [44]. One of the main challenges during FSW of Al-matrix composites is tool wear, due to the presence of hard reinforcing particles. Coating the tool, particularly when using high-speed steels, can reduce tool wear during the FSW of Al-matrix composites [45]. In the case of dissimilar welding between

Al composites and other Al alloys, it is recommended to offset the tool toward the Al alloy. This reduces tool wear by minimizing contact with the reinforcing particles in the Al composite [46]. The presence of particles decreases the thermal conductivity of the Al-matrix, causing a rise in the peak temperature during FSW [47]. The stirring action of the tool in the NZ refines and redistributes the reinforcing particles, which also promotes grain refinement. This equiaxed grain refinement occurs due to a mechanism called particle-stimulated nucleation (PSN). When the particles are larger than 0.5 μm, significant energy is stored around them during the deformation caused by FSW, providing many recrystallization sites. A high concentration of these particles leads to a fine, recrystallized grain structure. Additionally, the particles inhibit grain growth through the Zener drag mechanism [48]. This process results in a distinct hardness profile in the weld zone of Al-matrix composites compared to Al alloys [49]. As an example, the microhardness profile of a FSW joint between A390-SiC composite and AA2024-T4 shows an increase in hardness in the NZ compared to both base metals, due to the dispersion of fine SiC particles. A decrease in hardness is evident in the TMAZ of AA2024, whereas an increase is seen in the TMAZ of the cast composite Al alloy compared to its base metal. The reduced hardness in the TMAZ of AA2024 is due to the absence of reinforcing particles in this region [50]. While the TMAZ of the cast Al alloy lacks reinforcing particles, its hardness increases due to partial plastic deformation, which enhances hardness through a strain-hardening mechanism relative to the base metal having a cast microstructure. Figure 5.11a shows the schematic of the hardness profile in a cast Al alloy after the FSW process. The increase in hardness in the NZ is attributed to grain refinement as well as refinement of eutectic structures and primary phases like Si and Cu [51]. The increase in hardness in its TMAZ is due to work-hardening caused by the tool of FSW. The presence of particles causes an additional increase in this region, as mentioned. Figure 5.11b shows the schematic of the hardness profile in an HTA in the presence and absence of reinforcing particles.

The limiting grain size ($\overline{D}_{\max}$) in the presence of reinforcing particles during dynamic recrystallization is given by the following equation [52]:

$$\overline{D}_{\max} = \frac{4r}{v_f} \tag{5.5}$$

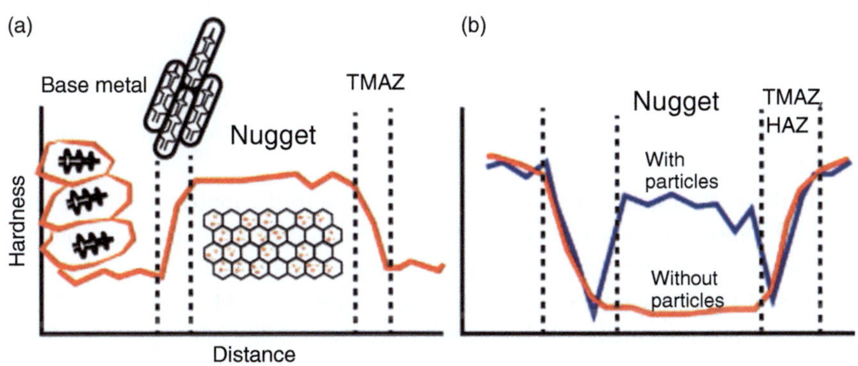

Figure 5.11 (a) Schematic of hardness profile in a cast Al alloy after FSW process. (b) Schematic of hardness profile in an HTA in the presence and absence of reinforcing particles.

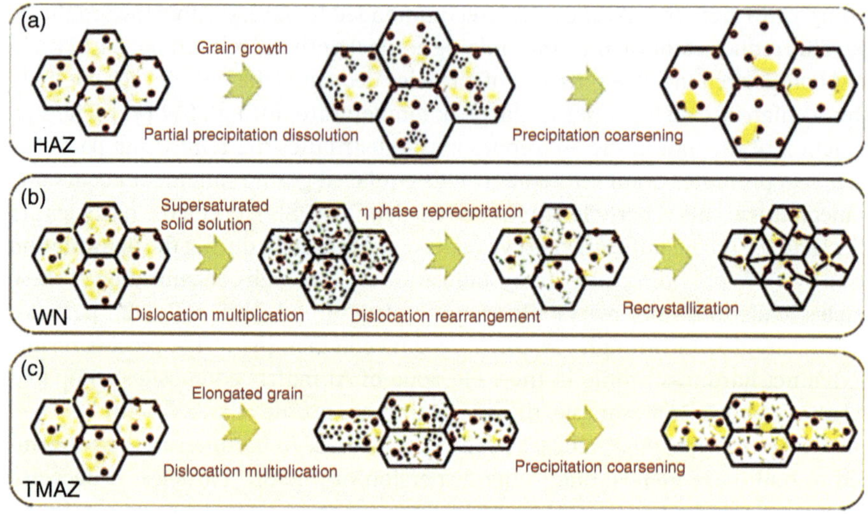

Figure 5.12 Schematic of microstructure evolution during FSW of Al-matrix composites during FSW in (a) HAZ, (b) NZ, and (c) TMAZ. Source: Ref. [53].

where r is the radius of the reinforcing particles and v_f is its volume fraction. In addition to grain refinement, the presence of reinforcements in the NZ enhances re-precipitation. This is due to the heterogeneous nucleation sites that form at the interface of the reinforcing particles [53]. A summary of microstructure evolution during FSW of Al-matrix composites is shown in Figure 5.12. In the HAZ (Figure 5.12a), coarsening and dissolution of precipitates are observed, along with limited grain growth. In the NZ (Figure 5.12b), recrystallization and significant grain refinement occur, aided by the reinforcing particles, along with precipitation at their interfaces. In the TMAZ (Figure 5.12c), an increase in dislocation density due to deformation is observed, as well as the coarsening of precipitates.

The dispersion of reinforcing particles in the NZ is not always homogeneous. A well-known structure in the NZ is the banding pattern, consisting of alternating bands rich and poor in precipitates [54]. When Al-matrix composites undergo FSW, the high concentration of reinforcing particles leads to the formation of reinforcement-rich bands. These bands negatively affect the mechanical properties of the joint due to the brittleness of the reinforcements, which can act as initiation sites for fractures in these regions. [55]. The primary advantage of reinforcing particles in Al-matrix composites is their ability to reduce the HAZ softening during welding [55].

During the FSW process, the particles may react with the matrix and form new phases [31]. One example is Al-matrix composites reinforced with carbon nanotubes (CNTs), where CNTs transform into Al_4C_3 during the FSW of CNT/Al—Cu—Mg composites [14]. The semi-coherent interface between the Al_4C_3 phase and the Al-matrix promotes strong bonding and enhances the stability of the particles. This also improves the mechanical properties of the NZ through the Orowan strengthening effect of the Al_4C_3 particles [56]. High-resolution TEM images of the interface

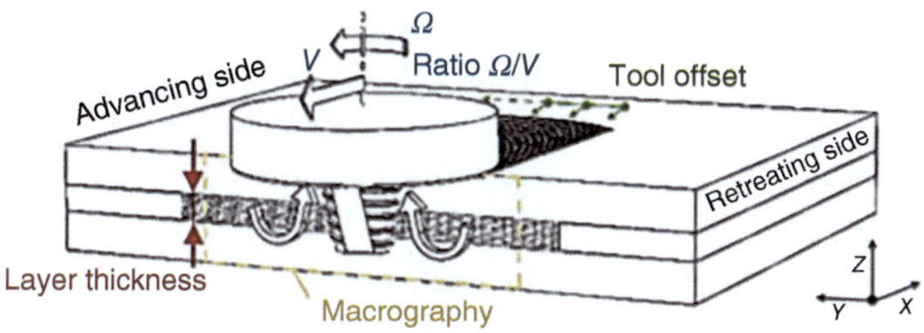

Figure 5.13 Process of upward FSP used to make Al-matrix composite by uniform dispersion of the particles. Source: Ref. [59].

between the Al_4C_3 and Al-matrix, along with the corresponding Fast Fourier Transform (FFT) pattern and the simulated pattern, indicate a semi-coherent interface, confirming the enhanced bonding at the interface.

While FSW is used to join Al composites, friction stir processing (FSP) can be employed to produce Al-matrix composites. In this method, reinforcing particles are placed in a groove, and the tool pin is inserted into the groove, rotating at a predetermined speed as it traverses along the groove. The stirring action of the tool disperses the particles in the stir zone, similar to the NZ in FSW of Al-matrix composites, thereby creating an Al-matrix composite. The dimensions of the groove are critical, as they determine the volume fraction of the reinforcing particles [57]. FSP is also commonly used to create surface composites that enhance the wear resistance of Al alloys [58]. However, the groove approach presents challenges, such as nonuniform particle dispersion during FSP. To address these issues, modifications have been developed. One such method is upward FSP, where reinforcing particles are placed between two stacked sheets [59]. In this technique, a threaded pin with a controlled pitch ratio is used to promote upward particle flow, ensuring uniform dispersion throughout the stir zone (Figure 5.13).

5.4 FSW of Al to Dissimilar Materials

FSW of Al to other materials needs special care to keep the temperature below the melting point of Al and avoid excessive mixing of two materials in the solid state, as these promote the formation of detrimental IMCs. The key point is to offset the pin of the tool into Al with a slight contact with the other material in a butt configuration, so as to keep the temperature as low as possible. This is because the peak temperature during FSW is below the melting point of the material with which the tool is in contact. The other important factor is to choose the right direction of the rotation of the tool. The flow direction of materials around the tool is from the advancing side to the retreating side. As most of the materials that are welded to Al have higher melting points than Al (steel, copper, titanium), it is recommended to put them on the advancing side, so the fragments of these materials can be easily dragged into Al, which is on the retreating side [60]. Doing vice versa causes the formation of defects on the advancing side, as the softened Al on the advancing side cannot be flown into the other material on the retreating side. This is because

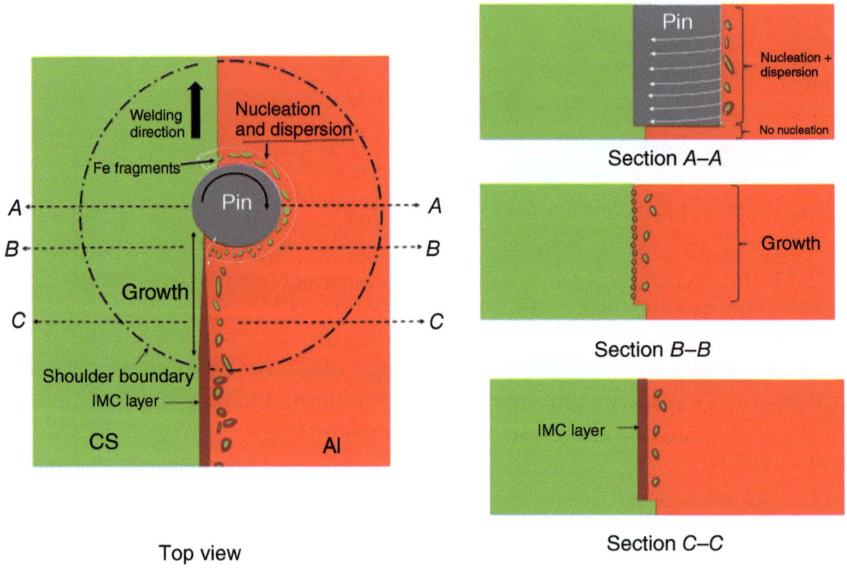

Figure 5.14 Schematic of the tool position during FSW of steel. Source: Ref. [63].

the temperature is below the melting point of Al, which is far below the softening temperature of the other material making it hard and resistant to allow the Al to flow into it [60]. Figure 5.14 presents the top view and side view of FSW of Al to steel, showing a schematic of material flow around the pin. However, in some cases the harder material may be put on the retreating side, which seems to contravene the previous point. As an example, during FSW of thin sheets of Al/Cu with 1 mm thicknesses, Al is placed on the advancing side [61]. The main reason behind this is that in thin sheets the material flow is mainly driven by the shoulder and not the pin [62]. In this case, the material flow is from the retreating side to the advancing side.

The offset of the tool into Al is not 100%, and therefore the other material on the advancing side is cut away and distributed into Al in front of the pin from the advancing side to the retreating side. This is necessary to make sure that the oxide layer on the other material is disrupted to offer a virgin surface for metallurgical bonding with Al. Al is forged behind the tool toward this activated surface through the material flow from the retreating side to the advancing side. When Al is put on the advancing side, Al is forged toward the inactivated surface, making a tunnel defect along with a kissing bond between the two materials [64]. The dispersion of the separated particles into Al makes a composite structure, which manifests as fluctuation in hardness profile in the NZ of Al [64]. The rubbing of the other material by the pin needs to be minimized to avoid large fragments separation. These large fragments inside Al-matrix interfere with the material flow, which is almost inside Al. Hence, these large fragments are accompanied by void defects, which degrade the performance of the joint [65]. Figure 5.15a shows a schematic of the microstructure evolution during FSW of Al to a dissimilar material (Cu in this figure). The corresponding hardness profile is shown in Figure 5.15b. Al is in cold-worked condition showing the elongated grains in the base material due to cold rolling. The fragments of Cu in the

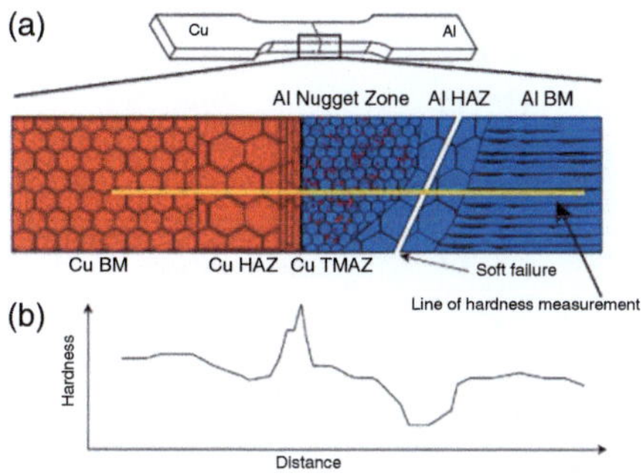

Figure 5.15 (a) Schematic of microstructure evolution during FSW of Al to Cu. (b) Corresponding hardness profile measured along the yellow line in (a). (c) Notched tensile specimen to characterize the joint strength. Source: Ref. [66].

NZ of Al produce a composite structure with a high hardness value. The most important feature of this figure is the absence of NZ in Cu, as the pin is offset into Al. Cu in the vicinity of the joint interface possesses a TMAZ that is influenced by the plastic deformation of the pin, which produces elongated grains with a high density of dislocations. The peak in hardness profile in this region is due to this phenomenon [66].

Usually, the tensile specimens prepared from the butt joints fail from the interface due to the presence of IMCs, which are often oriented perpendicular to the loading direction. This fracture propagates through the weakest location that depends on the IMC layer thickness, IMC type, presence of defects, residual stress, and morphology of the structures at the interface layer. It is well known that the thicker the IMC layer, the more brittle and weaker the joint becomes. [67]. Several approaches are used to control the thickness of IMC layer during FSW of Al to dissimilar materials. Controlling the FSW parameters and using a cooling medium are the most used approaches to do so [68]. When the joint strength is higher than the other regions, the failure occurs outside of the interface, most probably in the HAZ of the weaker material. In this state, to evaluate the interface strength, a notch is produced in the tensile specimen to concentrate stress at the interface, as can be seen in Figure 5.15c.

An example is provided in Figure 5.16, which shows the schematic of fracture of Al/steel joints welded by the FSW process in a butt configuration. This mechanism is obtained through a detailed analysis of the IMCs by TEM as well as a detailed analysis of the fracture surface studied by SEM [69]. Based on this study, the interface of the IMC layer of η and steel is the weakest location when loaded in the normal direction. These locations show a brittle feature when studied by SEM. When the loading condition changes from normal to shear, this interface is resistant and the failure occurs through steel by localized plastic deformation at the triple junction of the grain boundaries of η and steel (Figure 5.16d) [69].

The morphology of the IMC layer at the interface affects the fracture toughness through multiple deflections of the fracture path. Figure 5.17a and b shows the interface of Al/steel welded by FSW in the as-welded and heat-treated conditions, respectively. The protrusion of steel in the IMC layer in the as-welded condition is obvious. After heat treatment, these protrusions disappear and the thickness of the IMC layer increases from 1.8 to 3.5 μm. As these protrusions of steel are

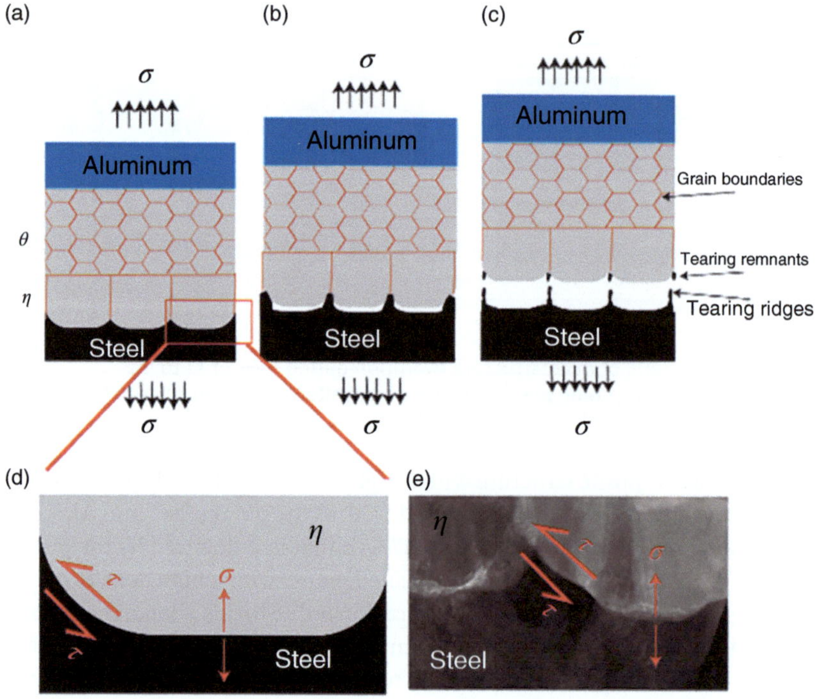

Figure 5.16 Fracture of the interface of Al/teel interface welded by FSW in butt configuration. (a) Two IMC phases, θ and η with fine and coarse grains at the interface. (b) Brittle fracture through the interface of η and steel where the stress state is in normal direction. (c) Ductile fracture at the triple junction of grain boundaries where the stress state is a mixed mode of shear and normal. (d, e) Schematic image of the stress state at the interface based on the TEM image. Source: Ref. [69]/with permission of Elsevier.

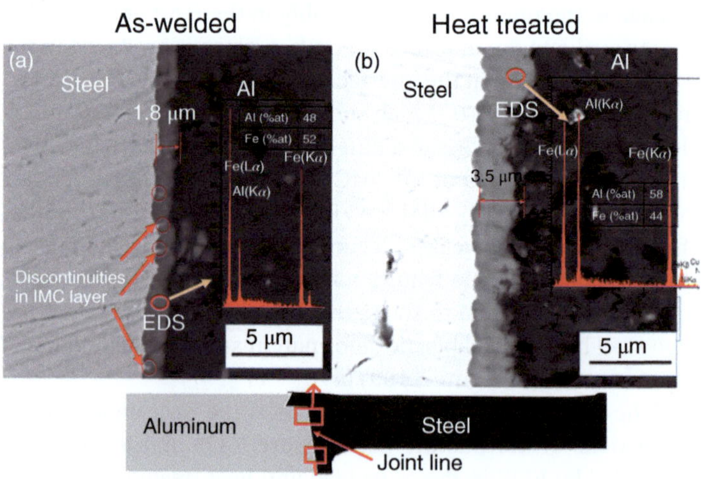

Figure 5.17 SEM images of the interface of Al/steel join made by FSW. (a) As-welded condition. (b) After PWHT. Source: Ref. [70]/Springer Nature.

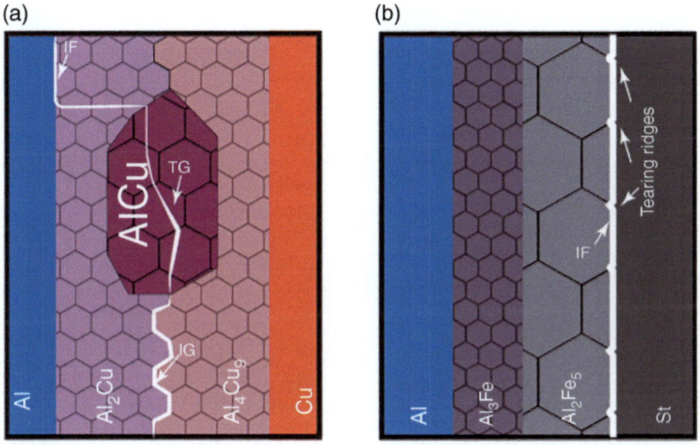

Figure 5.18 Schematic of fracture propagation path at the interface of the joints made by FSW. (a) Al/Cu joint. (b) Al/steel joint. Source: Ref. [66].

ductile, they act as a barrier to crack propagation and hence increase the toughness and strength of the joint. After PWHT, the IMC layer becomes uniform and thick and a fully ductile fracture occurs during tensile testing [70]. In the case of other dissimilar joining such as Al/Mg, the same behavior exists. Even a thin, uniform, and continuous IMC layer can be more detrimental to mechanical properties than a thicker layer where IMCs are dispersed within a soft matrix [71].

Besides morphology, the type of IMCs determines the fracture behavior. Various IMCs with different toughness influence the fracture path depending on their distribution and morphology. Figure 5.18a shows the schematic of IMCs at the interface of an Al/Cu joint made by FSW in a butt configuration. Detailed analyses using fractography and phase analysis reveal that the fracture path follows an inter-granular route between Al_2Cu and Al_4Cu_9. Crack deflection occurs when cracks reach AlCu phases distributed between Al_2Cu and Al_4Cu_9. This mechanism enhances joint toughness, despite the brittleness of the IMC responsible for the deflection. It is identified as the primary reason for the higher strength of Al/Cu joints compared to Al/steel joints, where a straight propagation path is observed [66].

When the joint design is lap, the mechanism of joining differs from butt welding. As a rule of thumb, Al as the lower melting material is always placed at the top of the joint to keep the temperature as low as possible. The height of the pin is slightly larger than the thickness of the upper material, so as to produce material flow and disperse the oxide layer from the surface of the bottom material. In contrast to similar welding of Al alloys in a lap configuration, where the hook formation degrades the joint strength due to cold lap formation, in dissimilar welding the hook formation has an anchoring effect that plays an important role in joint strength. In other words, the main mechanism of joining in FSW of dissimilar material to Al is mechanical interlocking [72]. The importance of mechanical interlocking is due to a lack of strong metallurgical bonding due to a massive IMC formation and the presence of defects. The new tool design and selection of optimum parameters to

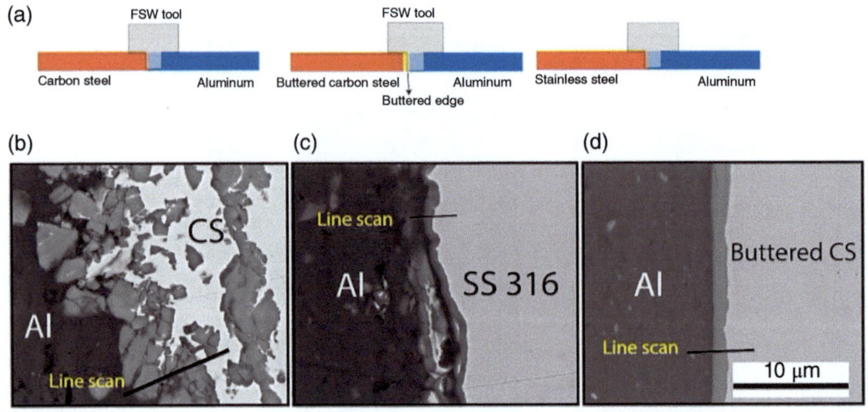

Figure 5.19 (a) Schematic of buttering stainless steel on the edge of carbon steel for FSW. SEM images of the interface of (b) Al/carbon steel, (c) Al/stainless steel, and (d) Al/buttered steel. Source: Ref. [79]/with permission of Elsevier.

obtain a lap weld free of defects and controlled thickness of IMCs layer fades away the importance of mechanical interlocking [73]. Conventional tools cause hook formation at the interface due to upward material flow. Suppression of this upward flow and turning it to a horizontal flow eliminates the hook formation and expands the bonding area where a metallurgical bond exists. Lap joints are evaluated by the shear tensile testing method, meaning that the stress state at the interface is shear. That is why the lap joints of dissimilar materials are less sensitive to the thickness of the IMCs layer, as already reported in several studies [74]. It is instead dependent on the bonding area, defect formation, the type of IMCs, and morphology of the weld interface [74–76]. The reports that highlight the importance of the IMCs layer thickness usually neglect these factors [77, 78].

The contribution of alloying elements in the formation of IMCs and their kinetics of growth is well addressed in the literature [63]. This gave a good insight to engineer the IMCs while FSW of Al to steel. A buttering technique is used to deposit stainless steel on the edge of carbon steel, preparing it to be welded to Al by FSW (Figure 5.19a). The presence of Ni and Cr in the butter layer helps to reduce the kinetics of growth of the IMC layer during FSW to Al [79]. Figure 5.19 shows SEM image of the interface of Al/carbon steel welded by FSW, where a massive and thick IMC layer is present. Figure 5.19c shows the interface of Al/stainless steel welded by FSW, where a thin IMC layer is observed. The same IMC layer is observed when buttered carbon steel is welded to Al, as can be seen in Figure 5.19d. An increase of 60% in tensile strength was observed by using this technique [79].

5.5 Joint Design

The most common joint configuration for joining Al alloys is butt joint. Lap configuration adds weight to the structure and therefore is less preferred than butt joint. Lap joint is more suited to joining dissimilar materials to Al, as it helps to make mechanical interlocking between the two materials at the interface. It is also used for joining thin sheets, as it is challenging to weld thin sheets in a butt configuration by FSW [80]. This is due to excessive thinning of the sheet resulting from the high force of the tool. Micro-FSW with a zero tilting angle of the tool helps to overcome thinning of the sheets during FSW [38]. The tilt angle of the tool along with a concavity in the shoulder is necessary during conventional FSW to provide enough pressure behind the shoulder for material flow. Another issue during FSW of thin Al sheets is related to the low stiffness of the sheet, which causes undulation of the sheet due to fixture pressure as well as rising of the sheet in front of the FSW tool [37]. So, clamping of the sheets during FSW needs to be as close as possible to the weld seam to mitigate the thickness reduction [37, 81].

Joining of thin sheets of Al (1 mm and lower) needs a high rotation speed (10,000–16,000 rpm), so a slight plunge depth of the pin into Al can complete the joint without thinning [82]. This high rotation speed of the tool causes the axial force of the tool to be reduced considerably and therefore robotic arms can be used for FSW. Robotic arms have low stiffness, and that is why high-speed FSW is suited to be mounted on robot arms [82]. At high rotation speeds, a cooling media is needed to avoid distortion during FSW. A saddle-shape distortion is produced during FSW, whose severity depends on the process parameters and the fixture pressure and design [37]. The process parameters intensify the distortion by increasing the heat input. For example, lower welding speed causes more severe distortion during FSW of thin Al sheets clamped tightly [37]. The peak temperature of high-speed FSW is higher than conventional FSW, but the exposure time to this high temperature is lower due to higher welding speed, producing lower distortion in high-speed FSW [83]. Table 5.1 summarizes the process parameters used for FSW of thin Al sheets along with the special care to lessen the issues related to thin sheets such as distortion.

Tee-joint is another design that is used for both similar welding and dissimilar welding. This joint type is used in the aircraft fuselage [41]. Figure 5.20 shows a

Table 5.1 Process parameters for FSW of thin Al sheets.

Thickness and type of Al	Tool rotation speed (rpm)	Welding speed (mm/min)	Diameter of the shoulder (mm)	Diameter of the pin (mm)	Special care	Reference
0.8, AA2024	400–1000	50–150	15	5 (cylindrical)		[84]
1, AA2014	10,000–16,500	100	6	1.2–1.5 (taper)	Underwater cooling	[82]
0.5, AA6061	2000	150–350	6.4	2.0–2.3 (taper)	Tight clamping	[37]
0.5, AA6061	1500	300–600	6	1.65–2.0 (taper)	0 tilt angle	[38]
0.8, AA6061	2000–8000	300–1500	6	2–3 (taper)	Helix on shoulder	[81]
0.8, AA6061	8000	1500	6	2–3 (taper)	Helix on shoulder	[85]

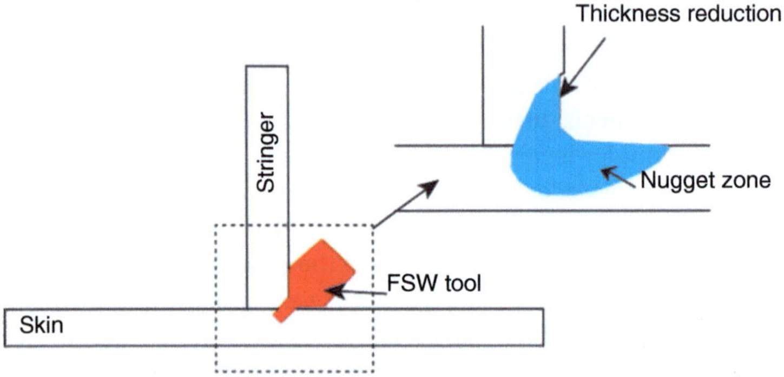

Figure 5.20 Schematic of FSW for T-joining of Al sheets along with the cross section of the joint.

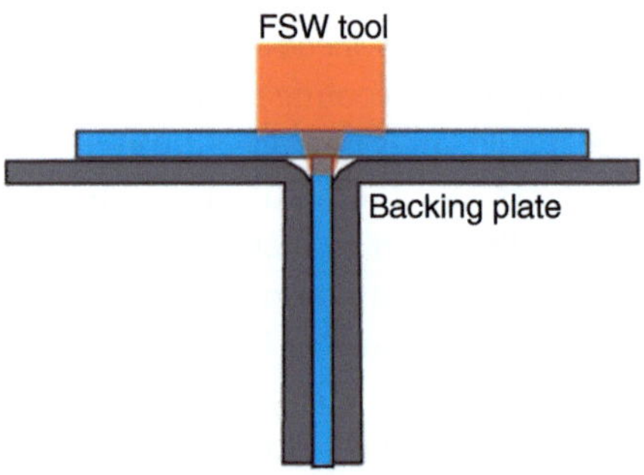

Figure 5.21 Schematic of Tee-joint design produced by FSW in lap configuration.

schematic for FSW of Al sheets in T-configuration where the stringer is placed on the advancing side. The shoulder is in contact with the two sheets. Due to the material flow from the advancing side to the retreating side, a thinning occurs in the stringer [86]. The material swept away from the advancing side is not fully transferred to the retreating side, remaining in the fillet area (Figure 5.20).

Tee-joint design can be produced by FSW in a lap configuration, as depicted in Figure 5.21. A free space between the backing plate and the corner is considered to make a fillet curvature through material flow to this cavity [87].

The most common defects that appear in this design are the kissing bond in the double corners and tunnel defect in the bottom (Figure 5.22). The kissing bond forms due to insufficient stirring of the tool and the consequent lack of shear deformation at the interface of skin/stringer far from the pin-influenced region [39]. The kissing bond degrades the mechanical performance of the joint when loaded in the skin direction, while the tunnel defect degrades it when loaded in the stringer direction [39]. The kissing bond increases the stress intensity factor at the corners during cyclic loading, leading to fatigue crack initiation [88]. In addition to mechanical performance, the corrosion resistance of the joint is also inflicted by kissing bonds. The crevice corrosion and galvanic corrosion are two types of corrosion that occur when welding similar and dissimilar Al alloys, respectively [89].

To reduce the defects obtained in a lap configuration, Tee-joints can be obtained by welding in a butt configuration, as shown in Figure 5.23, using a new design

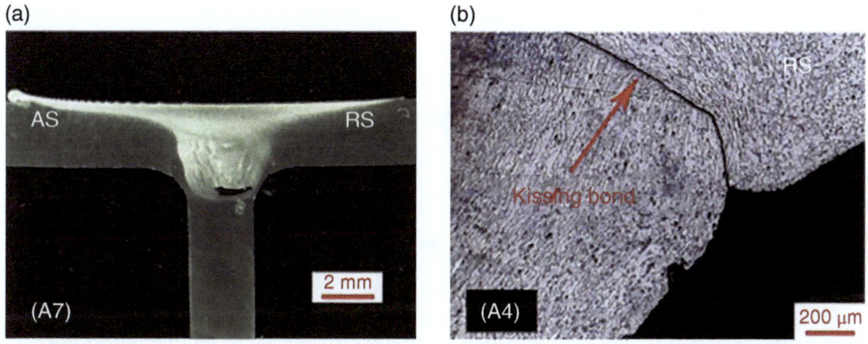

Figure 5.22 Common defects that form in Tee-joints, when welded in lap configuration according to Figure 5.20. (a) Tunnel defect. (b) Kissing bond. Source: Ref. [39]/with permission of Elsevier.

Figure 5.23 Tee-joints of Al alloys by FSW in butt configuration.

of the tool named progressive pyramidal pin [90]. This pin produces a larger NZ which extends to both interfaces and produces higher heat and shear force, helping to disrupt the oxide layer at the interfaces.

Butt welding of Al to dissimilar materials by FSW needs special care for tool geometry and dimension as well as the proper offset of the tool into Al. The best tool geometry to maintain a uniform rubbing of the dissimilar material, is cylindrical tool. The taper shape of the pin when inserted into the dissimilar material causes large fragments to be taken away into Al at the top which is accompanied by voids formation (Figure 5.24a). To avoid this defect, the tool needs to be offset more toward Al, making the least insert into the other material. This causes the bottom of the pin to stay away from the dissimilar material, so that at the bottom no contact exists between the pin and the dissimilar material (Figure 5.24b). This causes a lack of penetration at the bottom. The cylindrical pin establishes a uniform offset and contact with the dissimilar material, causing a uniform rubbing [65].

Dissimilar Tee-joining is more challenging than similar joining by FSW due to different materials properties and IMC formation. Tee-joining of Al/Ti in two different lap welding configurations, Ti as skin and Al as skin, yields different properties. Like lap joining, when putting Al on top (as skin) yields better results and the mechanical interlocking plays a major role in joining [91]. Modification in

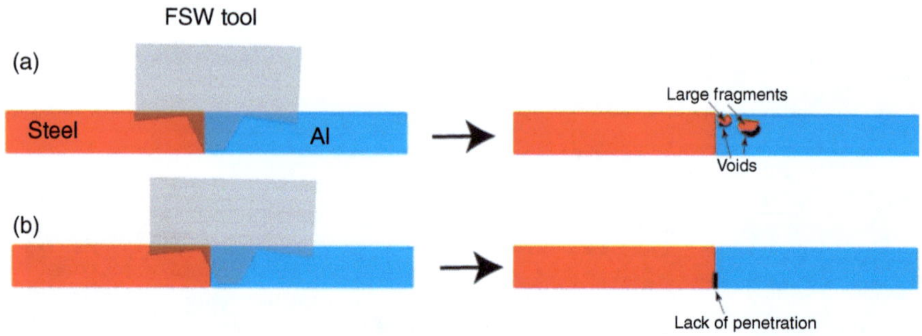

Figure 5.24 Schematic showing the geometry of the pin and the tool offset effect during FSW of Al to a dissimilar material. (a) An excess insert of the pin into dissimilar material at top. (b) Sufficient insert of the pin into dissimilar material at top, but no insert at the bottom.

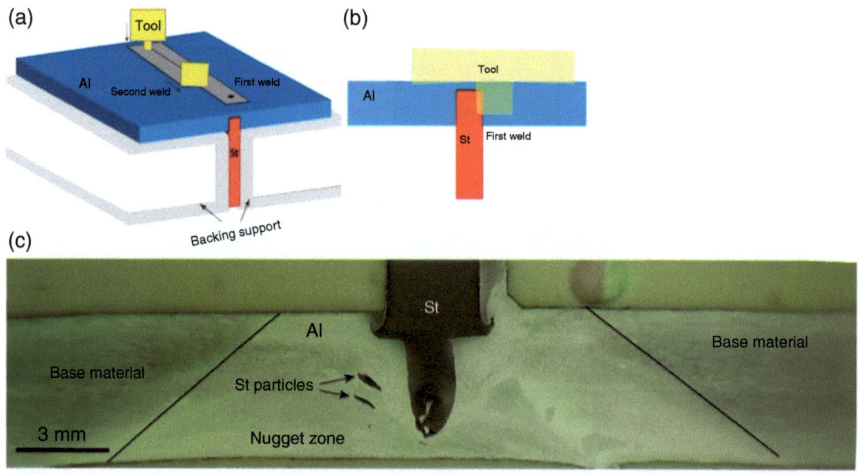

Figure 5.25 (a and b) Schematic of Al/steel Tee-joint design. (c) Resulting weld cross section. Source: Ref. [92]/MDPI/CC BY 4.0.

the design of the weld can make metallurgical bonding as depicted in Figure 5.25. This design comprises both lap and butt weld whose weld cross section is shown in Figure 5.25c. The mechanical behavior of this joint is highly influenced by the loading direction with respect to the interfaces. The best performance of the joint is obtained when the interface is parallel to the applying load, which makes shear stress on the IMC layer [92].

5.6 Simulation of the Process

Like fusion welding processes, FSW can be simulated by thermal-mechanical model to obtain temperature history, residual stress, and distortion. In this case, the tool of FSW is modeled as a heat source. This method does not yield the material flow and the history of plastic deformation induced by the rotation of the tool. One important goal of simulation of the FSW process is to study the material flow in order to investigate the mechanism of defects formation such as tunneling defects.

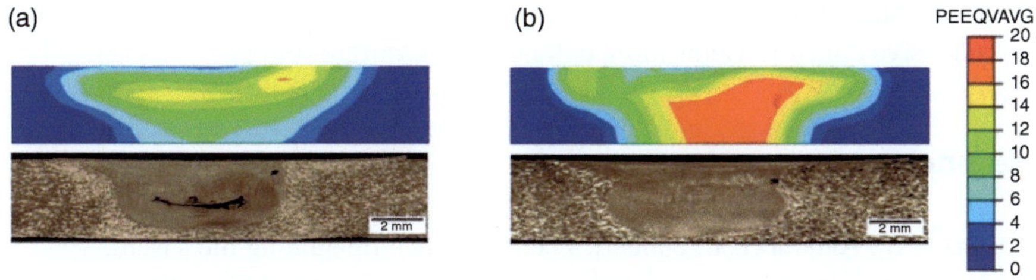

Figure 5.26 Equivalent plastic strain map obtained during FSW of AA6082 obtained at two rotation speeds of the tool along with the real cross section of the welds obtained via experiment. (a) Insufficient material flow. (b) Sufficient material flow. Source: Ref. [94]/with permission of Elsevier.

Arbitrary Lagrangian–Eulerian (ALE) or coupled Eulerian–Lagrangian (CEL) FE models are used to simulate the material flow during FSW. In the CEL method, the tool is considered a rigid Lagrangian body and the workpiece is simulated with Eulerian formulation using a contact definition between the two [93]. More realistic results are obtained using these approaches to analyze the FSW process. Obtained equivalent plastic strain in the stir zone of the weld is a good indicator of the defects, such as tunneling, which are caused by insufficient material flow (Figure 5.26) [94].

The tool/workpiece contact conditions determine the temperature and friction behavior, which affect the material flow. The more accurate this friction coefficient is, the more realistic the temperature field and material flow become. The friction coefficient is temperature-dependent, and taking this into account makes the simulation results more reliable [95]. The material flow characteristics depend on the temperature and strain rate, which is quantitatively described by the following equation [96]:

$$\sigma = B\dot{\varepsilon}^m \tag{5.6}$$

where σ is the flow stress, $\dot{\varepsilon}$ is the strain rate, m is the strain rate sensitivity, and B is a parameter that depends on temperature. The coefficients of this equation are obtained through quasi-static compression tests at various temperatures and strain rates. For some alloys such as AA5083, the strain rate sensitivity is high, and therefore the material flow stress near the tool, where the strain rate is high, is very large, even though the temperature is very high. This makes the material flow difficult, making these alloys prone to tunnel defect [96]. Elastic–plastic Johnson-Cook law is an extended formula that considers all the influencing factors on flow stress of Al [87]:

$$\sigma = (A + B\varepsilon^n)\left[1 + C\ln\left(1 + \frac{\dot{\varepsilon}}{\dot{\varepsilon}_0}\right)\right]\left[1 - \left(\frac{T - T_{room}}{T_{melt} - T_{room}}\right)^m\right] \tag{5.7}$$

where A, B, and C are constants related to the material. $\dot{\varepsilon}$ is the strain rate, $\dot{\varepsilon}_0$ is the reference strain rate, T_{room} is the reference temperature, and T_{melt} is the melting point.

Choosing the right approach to simulate the FSW process depends on the desired outcomes, as well as the cost and time constraints of the simulation. While uncoupled thermal-mechanical simulation methods are faster and require less computing power, the CEL method demands significantly more time and a much stronger computer for calculations. However, the CEL method provides more comprehensive data and yields more detailed and accurate results.

References

1 Di Bella, G., Alderucci, T., Salmeri, F. et al. (2022). Integrating the sustainability aspects into the risk analysis for the manufacturing of dissimilar aluminium/steel friction stir welded single lap joints used in marine applications through a life cycle assessment. *Sustainable Futures* 4: 100101.

2 De Filippis, L.A.C., Serio, L.M., Palumbo, D. et al. (2017). Optimization and characterization of the friction stir welded sheets of AA 5754-H111: monitoring of the quality of joints with thermographic techniques. *Materials (Basel)* 10 (10): 1165.

3 Qiao, Q., Su, Y., Cao, H. et al. (2020). Effect of post-weld heat treatment on double-sided friction stir welded joint of 120 mm ultra-thick SiCp/Al composite plates. *Materials Characterization* 169: 110668.

4 Rathinasuriyan, C., Senthil Kumar, V.S., Savitha, V. et al. (2022). Mechanical and metallurgical properties of GTAW, GMAW and FSW lap joints on AA6061-T6 alloy. *Advances in Materials and Processing Technologies* 8 (3): 3231–3247.

5 Kotari, S., Punna, E., Gangadhar Reddy, S.M. et al. (2020). Experimental investigation and micro structural characteristics of GTAW and FSW welded dissimilar Aluminum alloys. *IOP Conference Series: Materials Science and Engineering*, Hyderabad, India (27–28 August 2020). IOP Publishing.

6 Qiao, Y., Zhang, H., Zhao, L. et al. (2020). Fatigue crack growth properties of AA 5754 aluminum alloy gas tungsten arc welding and friction stir welding joints. *Journal of Materials Engineering and Performance* 29: 2113–2124.

7 Ahmed, M.M., El-Sayed, M.M., Fydrych, D. et al. (2023). Friction stir welding of aluminum in the aerospace industry: the current progress and state-of-the-art review. *Materials* 16 (8): 2971.

8 Beygi, R., Galvão, I., Nematzadeh, F. et al. (2023). A novel procedure for friction stir welding aluminium to copper by using an aluminium run-on plate. *Materials Science Forum* 1110: 83–90. Trans Tech Publication.

9 Torabi, K., Beygi, R., Eisaabadi, G. et al. (2023). The effect of tool rotation speed on the formation of eutectic structure during friction stir welding of aluminum to magnesium. *Applied Sciences* 13 (12): 7133.

10 Kasaei, M., Beygi, R., Carbas, R.J.C. et al. (2023). A review on mechanical and metallurgical joining by plastic deformation. Discover. *Mechanical Engineering* 2 (1): 5.

11 Nonnenmann, T., Beygi, R., Carbas, R.J.C. et al. (2023). Feasibility study on hybrid weld-bonding between dissimilar material for automotive industry. *International Journal of Adhesion and Adhesives* 121: 103316.

12 Steuwer, A., Dumont, M., Altenkrich, J. et al. (2011). A combined approach to microstructure mapping of an Al–Li AA2199 friction stir weld. *Acta Materialia* 59 (8): 3002–3011.

13 Prabhu, R.S.B., Shettigar, A.K., Manjunath Patel, G.C. et al. (2022). Influence of process variables on joint attributes of friction stir welded aluminium matrix composite. *Advances in Materials and Processing Technologies* 8 (2): 1550–1559.

14 Li, X., Zhang, Z., Peng, Y. et al. (2022). Microstructure and mechanical properties of underwater friction stir welding of CNT/Al–Cu–Mg composites. *Journal of Materials Research and Technology* 18: 405–415.

15 Davis, J. (1993). *Aluminum and Aluminum Alloys*. ASM International.

16 Yu, C., Ding, H., Li, J.z. et al. (2015). Effect of welding heat input and post-welded heat treatment on hardness of stir zone for friction stir-welded 2024-T3 aluminum alloy. *Transactions of Nonferrous Metals Society of China* 25 (8): 2524–2532.

17 Zhang, J., Feng, X.S., Gao, J.S. et al. (2018). Effects of welding parameters and post-heat treatment on mechanical properties of friction stir welded AA2195-T8 Al–Li alloy. *Journal of Materials Science & Technology* 34 (1): 219–227.

18 Kosturek, R., Śnieżek, L., Wachowski, M. et al. (2019). The influence of post-weld heat treatment on the microstructure and fatigue properties of Sc-modified AA2519 friction stir-welded joint. *Materials (Basel)* 12 (4): 583.

19 El-Danaf, E.A. and El-Rayes, M.M. (2013). Microstructure and mechanical properties of friction stir welded 6082 AA in as welded and post weld heat treated conditions. *Materials & Design* 46: 561–572.

20 İpekoğlu, G., Erim, S., Çam, G. et al. (2014). Investigation into the influence of post-weld heat treatment on the friction stir welded AA6061 Al-alloy plates with different temper conditions. *Metallurgical and Materials Transactions A* 45: 864–877.

21 İpekoğlu, G., Erim, S., and Çam, G. (2014). Effects of temper condition and post weld heat treatment on the microstructure and mechanical properties of friction stir butt-welded AA7075 Al alloy plates. *The International Journal of Advanced Manufacturing Technology* 70: 201–213.

22 Dong, J., Zhang, D., Zhang, W. et al. (2023). Effect of post-weld heat treatments on the microstructure and mechanical properties of underwater friction stir welded joints of 7003-T4/6060-T4 aluminium alloys. *Materials Science and Engineering: A* 862: 144423.

23 Sharma, C., Dwivedi, D.K., and Kumar, P. (2013). Effect of post weld heat treatments on microstructure and mechanical properties of friction stir welded joints of Al–Zn–Mg alloy AA7039. *Materials & Design* 43: 134–143.

24 Sajadifar, S.V., Moeini, G., Scharifi, E. et al. (2019). On the effect of quenching on postweld heat treatment of friction-stir-welded aluminum 7075 alloy. *Journal of Materials Engineering and Performance* 28: 5255–5265.

25 Zhang, C., Huang, G., Zhang, D. et al. (2020). Microstructure and mechanical properties in dissimilar friction stir welded AA2024/7075 joints at high heat input: effect of post-weld heat treatment. *Journal of Materials Research and Technology* 9 (6): 14771–14782.

26 Lezaack, M.B. and Simar, A. (2021). Avoiding abnormal grain growth in thick 7XXX aluminium alloy friction stir welds during T6 post heat treatments. *Materials Science and Engineering A* 807: 140901.

27 Porter, D.A., Easterling, K.E., and Sherif, M.Y. (2021). *Phase Transformations in Metals and Alloys*. CRC Press.

28 Cayless, R. (1990). Alloy and temper designation systems for aluminum and aluminum alloys. In: *ASM Handbook Volume 2: Properties and Selection: Nonferrous Alloys and Special-Purpose Materials* (ed. ASM International). ASM International.

29 Rao, D., Huber, K., Heerens, J. et al. (2013). Asymmetric mechanical properties and tensile behaviour prediction of aluminium alloy 5083 friction stir welding joints. *Materials Science and Engineering: A* 565: 44–50.

30 Mironov, S., Inagaki, K., Sato, Y.S. et al. (2015). Effect of welding temperature on microstructure of friction-stir welded aluminum alloy 1050. *Metallurgical and Materials Transactions A* 46: 783–790.

31 Beygi, R., Mehrizi, M.Z., and Eisaabadi, G.B. (2017). Friction stir processing of al with mechanically alloyed Al–TiO_2–graphite powder: microstructure and mechanical properties. *Journal of Materials Engineering and Performance* 26: 1455–1462.

32 Saravanakumar, R., Rajasekaran, T., and Pandey, C. (2023). Underwater friction stir welded armour grade AA5083 aluminum alloys: experimental ballistic performance and corrosion investigation. *Journal of Materials Engineering and Performance* 32 (22): 10175–10190.

33 Tan, Y., Wang, X.M., Ma, M. et al. (2017). A study on microstructure and mechanical properties of AA 3003 aluminum alloy joints by underwater friction stir welding. *Materials Characterization* 127: 41–52.

34 Abnar, B., Gashtiazar, S., and Javidani, M. (2023). Friction stir welding of non-heat treatable Al alloys: challenges and improvements opportunities. *Crystals* 13 (4): 576.

35 Guo, N., Fu, Y., Wang, Y. et al. (2017). Microstructure and mechanical properties in friction stir welded 5A06 aluminum alloy thick plate. *Materials & Design* 113: 273–283.

36 Kumar, R., Upadhay, V., and Pandey, C. (2023). Effect of post-weld heat treatments on microstructure and mechanical properties of friction stir welding joints of AA2014 and AA7075. *Journal of Materials Engineering and Performance* 32 (24): 10989–10999.

37 Ahmed, S. and Saha, P. (2018). Development and testing of fixtures for friction stir welding of thin aluminium sheets. *Journal of Materials Processing Technology* 252: 242–248.

38 Huang, Y., Meng, X., Lv, Z. et al. (2019). Microstructures and mechanical properties of micro friction stir welding (μFSW) of 6061-T4 aluminum alloy. *Journal of Materials Research and Technology* 8 (1): 1084–1091.

39 Zhao, Y., Zhou, L., Wang, Q. et al. (2014). Defects and tensile properties of 6013 aluminum alloy T-joints by friction stir welding. *Materials & Design* 57: 146–155.

40 Beygi, R., Mehrizi, M.Z., Akhavan-Safar, A. et al. (2023). A parametric study on the effect of FSW Parameters and the tool geometry on the tensile strength of AA2024–AA7075 joints: microstructure and fracture. *Lubricants* 11 (2): 59.

41 Burek, R., Wydrzyński, D., Kubit, A. et al. (2021). The influence of the shoulder depth on the properties of the thin sheet joint made by FSW technology. *Aircraft Engineering and Aerospace Technology* 93 (1): 120–126.

42 Li, Z., Yue, Y., Ji, S. et al. (2016). Joint features and mechanical properties of friction stir lap welded alclad 2024 aluminum alloy assisted by external stationary shoulder. *Materials & Design* 90: 238–247.

43 Abolusoro, O.P., Khoathane, M.C., Mhike, W. et al. (2024). Influence of welding parameters and post weld heat treatment on mechanical, microstructures and corrosion behaviour of friction stir welded aluminium alloys. *Journal of Materials Research and Technology* 32: 634–648.

44 Cao, R., Kai, X., Qian, W. et al. (2024). Effect of in-situ ZrB2 nanoparticles on microstructure and mechanical properties of friction stir welding joints in 7N01 matrix composites. *Materials Characterization* 207: 113611.

45 Zuo, L., Shao, W., Zhang, X. et al. (2022). Investigation on tool wear in friction stir welding of SiCp/Al composites. *Wear* 498: 204331.

46 Wang, C., Wang, B., Xue, P. et al. (2022). Effect of tool offsetting on friction stir welding of dissimilar aluminium matrix composite and aluminium alloy. *Science and Technology of Welding and Joining* 27 (7): 586–593.

47 Senthilraj, K. and Rajamurugan, G. (2024). Effect of Al_2TiO_5 particle on the temperature distribution of friction stir welded dissimilar aluminium composites using response surface methodology. *Transactions of the Indian Institute of Metals* 77 (4): 1205–1215.

48 Imam, M., Sun, Y.F., Fujii, H. et al. (2018). Deformation characteristics and microstructural evolution in friction stir welding of thick 5083 aluminum alloy. *The International Journal of Advanced Manufacturing Technology* 99: 663–681.

49 Kumar, B.A., Dinaharan, I., and Murugan, N. (2022). Microstructural, mechanical and wear properties of friction stir welded AA6061/AlNp composite joints. *Journal of Materials Engineering and Performance* 31 (1): 651–666.

50 Aval, H.J. and Galvão, I. (2024). Characterization of friction stir welded Al–4Cu–Mg alloy/Al–16Si–4Cu–10SiC composite joint. *Journal of Advanced Joining Processes* 9: 100192.

51 Pan, Y. and Lados, D. (2017). Friction stir welding in wrought and cast aluminum alloys: weld quality evaluation and effects of processing parameters on microstructure and mechanical properties. *Metallurgical and Materials Transactions A* 48: 1708–1726.

52 Singh, T., Tiwari, S.K., and Shukla, D.K. (2022). Novel method of nanoparticle addition for friction stir welding of aluminium alloy. *Advances in Materials and Processing Technologies* 8 (1): 1160–1172.

53 Sun, K., Kai, X., Du, C. et al. (2024). Influence of ZrB_2 and Al_2O_3 nanoparticles on microstructure and mechanical property of friction stir welded joints of in-situ 7085Al matrix composites. *Journal of Materials Processing Technology* 326: 118323.

54 Beygi, R., Kazeminezhad, M., and Kokabi, A. (2012). Butt joining of Al–Cu bilayer sheet through friction stir welding. *Transactions of Nonferrous Metals Society of China* 22 (12): 2925–2929.

55 Miao, C., Du, C., Kai, X. et al. (2022). Microstructure and mechanical properties of friction stir welding joint of in-situ ZrB2/AA7085 composites. *Applied Surface Science* 586: 152740.

56 Li, X., Zhang, Z., Wang, Y. et al. (2022). Microstructure evolution and mechanical properties of CNT/Al composite joints via friction stir welding. *Science and Technology of Welding and Joining* 27 (8): 664–671.

57 Kesharwani, R., Jha, K.K., Imam, M. et al. (2024). Selection of groove dimension in the fabrication of particulate reinforced aluminium matrix composites using friction stir welding. *Materials Letters* 367: 136640.

58 Boopathi, S. (2022). An investigation on friction stir processing of Aluminum alloy-boron carbide surface composite. In: *Advances in Processing of Lightweight Metal Alloys and Composites: Microstructural Characterization and Property Correlation* (ed. V.V. Ramalingam, P. Ramasamy, and M. Govindaraju), 249–257. Springer.

59 Vidal, C., Ferreira, P.M., Inácio, P.L. et al. (2023). Particles' distribution enhancing in aluminum-based composites produced by upward friction stir processing. *The International Journal of Advanced Manufacturing Technology* 127 (5): 2745–2757.

60 Kulkarni, N., Mishra, R.S., and Yuan, W. (2015). *Friction Stir Welding of Dissimilar Alloys and Materials.* Butterworth-Heinemann.

61 Mao, Y., Ni, Y., Xiao, X. et al. (2020). Microstructural characterization and mechanical properties of micro friction stir welded dissimilar Al/Cu ultra-thin sheets. *Journal of Manufacturing Processes* 60: 356–365.

62 Galvão, I., Leal, R., Loureiro, A. et al. (2010). Material flow in heterogeneous friction stir welding of aluminium and copper thin sheets. *Science and Technology of Welding & Joining* 15 (8): 654–660.

63 Beygi, R., Carbas, R.J.C., Queiros, A. et al. (2022). Comparative study between stainless steel and carbon steel during dissimilar friction stir welding with aluminum: kinetics of Al–Fe intermetallic growth. *Metals and Materials International* 28 (8): 1–12.

64 Yazdipour, A. and Heidarzadeh, A. (2016). Effect of friction stir welding on microstructure and mechanical properties of dissimilar Al 5083-H321 and 316L stainless steel alloy joints. *Journal of Alloys and Compounds* 680: 595–603.

65 Kaushik, P. and Dwivedi, D.K. (2021). Effect of tool geometry in dissimilar Al-steel friction stir welding. *Journal of Manufacturing Processes* 68: 198–208.

66 Beygi, R., Carbas, R.J.C., Marques, E.A.S. et al. (2024). Mechanism of toughness enhancement of brittle fracture by intermittent η-intermetallic in Al/Cu joint made by FSW. *Materials Science and Engineering: A* 890: 145907.

67 Mehta, K.P. (2019). A review on friction-based joining of dissimilar aluminum–steel joints. *Journal of Materials Research* 34 (1): 78–96.

68 Eyvazian, A., Hamouda, A., Tarlochan, F. et al. (2020). Simulation and experimental study of underwater dissimilar friction-stir welding between aluminium and steel. *Journal of Materials Research and Technology* 9 (3): 3767–3781.

69 Beygi, R., Carbas, R.J.C., Barbosa, A.Q. et al. (2021). A comprehensive analysis of a pseudo-brittle fracture at the interface of intermetallic of η and steel in aluminum/steel joints made by FSW: microstructure and fracture behavior. *Materials Science and Engineering: A* 824: 141812.

70 Beygi, R., Mehrizi, M.Z., Akhavan-Safar, A. et al. (2021). Design of friction stir welding for butt joining of aluminum to steel of dissimilar thickness: heat treatment and fracture behavior. *The International Journal of Advanced Manufacturing Technology* 112: 1951–1964.

71 Beygi, R., Pouraliakbar, H., Torabi, K. et al. (2021). The inhibitory effect of stir zone liquefaction and eutectic-phase formation on the growth of γ/β intermetallics during dissimilar FSW of Al/Mg alloys. *Journal of Manufacturing Processes* 70: 152–162.

72 Safeen, M.W., Spena, P.R., Buffa, G. et al. (2020). Effect of position and force tool control in friction stir welding of dissimilar aluminum-steel lap joints for automotive applications. *Advances in Manufacturing* 8: 59–71.

73 Liu, J., Zilin, H., Xie, Y. et al. (2022). Interface stability and fracture mechanism of Al/Steel friction stir lap joints by novel designed tool. *Journal of Materials Processing Technology* 300: 117425.

74 Ho, H.H., Huy, H.H., Quan, N.M. et al. (2023). Mechanical performance of dissimilar friction stir welded lap-joint between aluminium alloy 6061 and 316 stainless steel. *Welding International* 37 (2): 101–110.

75 Liu, T., Gao, S., Ye, W. et al. (2023). Achievement of high-quality joints and regulation of intermetallic compounds in ultrasonic vibration enhanced friction stir lap welding of aluminum/steel. *Journal of Materials Research and Technology* 25: 5096–5109.

76 Kar, A., Vicharapu, B., Morisada, Y. et al. (2020). Elucidation of interfacial microstructure and properties in friction stir lap welding of aluminium alloy and mild steel. *Materials Characterization* 168: 110572.

77 Wang, X., Morisada, Y., Matsushita, M. et al. (2023). Interface structure and mechanical properties of Fe/Al dissimilar lap joints formed by friction stir welding using an adjustable tool. *Science and Technology of Welding & Joining* 28 (8): 701–709.

78 Geng, P., Morimura, M. et al. (2022). Elucidation of intermetallic compounds and mechanical properties of dissimilar friction stir lap welded 5052 Al alloy and DP590 steel. *Journal of Alloys and Compounds* 906: 164381.

79 Beygi, R., Carbas, R.J.C., Barbosa, A. et al. (2023). Buttering for FSW: enhancing the fracture toughness of Al–Fe intermetallics through nanocrystallinity and suppressing their growth. *Journal of Manufacturing Processes* 90: 233–241.

80 Lader, S.K., Baruah, M., and Ballav, R. (2023). Significance of underwater friction stir welding on the weld integrity of thin sheets of aluminum (AA1050-O) and brass (CuZn34) joints. *Materials Science and Engineering: A* 865: 144627.

81 Liu, F., Sun, Z.Y., Tuo, Y.F. et al. (2020). Effect of shoulder geometry and clamping on microstructure evolution and mechanical properties of ultra-thin friction stir-welded Al6061-T6 plates. *The International Journal of Advanced Manufacturing Technology* 106: 1465–1476.

82 Chen, S., Zhou, Y., Junrong, X. et al. (2017). High rotation speed friction stir welding for 2014 aluminum alloy thin sheets. *Journal of Materials Engineering and Performance* 26: 1337–1345.

83 Ni, Y., Liu, Y., Zhang, P. et al. (2022). Thermal cycles, microstructures and mechanical properties of AA7075-T6 ultrathin sheet joints produced by high speed friction stir welding. *Materials Characterization* 187: 111873.

84 Yue, Y., Wang, G., Yang, K. et al. (2018). Friction stir butt welding thin aluminum alloy sheets. *The International Journal of Advanced Manufacturing Technology* 96: 3139–3147.

85 Liu, F., Fu, L., and Chen, H. (2018). High speed friction stir welding of ultra-thin AA6061-T6 sheets using different backing plates. *Journal of Manufacturing Processes* 33: 219–227.

86 Furkon, W.M., Kim, D., Song, M. et al. (2022). Effect of post–weld heat treatment on multi-scale microstructures and mechanical properties of friction stir welded T-joints of Al–Mg–Si alloys. *Journal of Materials Research and Technology* 18: 496–507.

87 Salloomi, K.N. (2019). Fully coupled thermomechanical simulation of friction stir welding of aluminum 6061-T6 alloy T-joint. *Journal of Manufacturing Processes* 45: 746–754.

88 Duong, H.D., Okazaki, M., and Tran, T.H. (2021). Fatigue behavior of dissimilar friction stir welded T-lap joints between AA5083 and AA7075. *International Journal of Fatigue* 145: 106090.

89 Astarita, A., Squillace, A., Scala, A. et al. (2012). On the critical technological issues of friction stir welding T-joints of dissimilar aluminum alloys. *Journal of Materials Engineering and Performance* 21: 1763–1771.

90 Jesus, J., Costa, J.M., Loureiro, A. et al. (2018). Assessment of friction stir welding aluminium T-joints. *Journal of Materials Processing Technology* 255: 387–399.

91 Rana, H., Campanella, D., Buffa, G. et al. (2023). Dissimilar titanium-aluminum skin-stringer joints by FSW: process mechanics and performance. *Materials and Manufacturing Processes* 38 (4): 471–484.

92 Beygi, R., Talkhabi, A.A., Mehrizi, M.Z. et al. (2023). A novel lap-Butt joint design for FSW of aluminum to steel in tee-configuration: joining mechanism, intermetallic formation, and fracture behavior. *Metals* 13 (6): 1027.

93 Mishin, V., Shishov, I., Kalinenko, A. et al. (2022). Numerical simulation of the thermo-mechanical behavior of 6061 aluminum alloy during friction-stir welding. *Journal of Manufacturing and Materials Processing* 6 (4): 68.

94 Salih, O.S., Ou, H., and Sun, W. (2023). Heat generation, plastic deformation and residual stresses in friction stir welding of aluminium alloy. *International Journal of Mechanical Sciences* 238: 107827.

95 Liu, Q., Li, W., Zhu, L. et al. (2022). Temperature-dependent friction coefficient and its effect on modeling friction stir welding for aluminum alloys. *Journal of Manufacturing Processes* 84: 1054–1063.

96 Nakamura, T., Obikawa, T., Nishizaki, I. et al. (2018). Friction stir welding of non-heat-treatable high-strength alloy 5083-O. *Metals* 8 (4): 208.

Index

Welding Metallurgy of Aluminium Alloys: Design, Processes, and Simulations, First Edition.
Reza Beygi, Eduardo André de Sousa Marques, Ricardo João Camilo Carbas, and Lucas Filipe Martins da Silva.